AF725305

Texas STAAR Grade 4 Math Made Easy

Study Guide with Key Concepts, Review & Practice

Dr. A. Nazari

Copyright © 2026 Dr. A. Nazari

Published by View Math Education

ViewMath.com

All rights reserved. No part of this publication may be reproduced, distributed, or transmitted in any form or by any means, including photocopying, recording, or other electronic or mechanical methods, without the prior written permission of the author, except in the case of brief quotations embodied in critical reviews and certain other noncommercial uses permitted by copyright law, including Section 107 or 108 of the 1976 United States Copyright Act.

The information in this book is distributed on an "as is" basis, without warranty. While every precaution has been taken in the preparation of this work, neither the author nor the publisher shall have any liability to any person or entity with respect to any loss or damage caused or alleged to be caused directly or indirectly by the information contained in this book.

Copyright © 2026

YOUR QUICK REFERENCE

Grade 4 Math Made Easy

Hey there, math learner! ⭐

*This study guide gives you the **core ideas**, **clear examples**, and **practice problems** for every Grade 4 math topic — all in one compact book.*

Need a refresher? Flip to any topic.

Want to practice? Every section has problems with answers.

Let's dive in! 🚀

Read

Practice

Master

How to Use This Book

Each topic follows the same simple pattern.

Key Concept

The main idea, explained simply. Understand this and the rest clicks into place.

Worked Examples

Step-by-step solutions that show you exactly how it's done. Try covering the answer first!

Practice Problems

Your turn to solve! Every problem has an answer in the back of the book.

Owlbert's Tips

Memory tricks, shortcuts, and "watch out" warnings from your study buddy Owlbert.

Study Tip: *Focus on one topic at a time. Understanding beats rushing — take the time you need!*

X^1 Math Symbols You Should Know X^1

Keep this page handy as you work through the book!

Symbol	Name	What It Means	
$+$	*Plus (Add)*	*Combine amounts.*	$4{,}521 + 3{,}089 = 7{,}610$
$-$	*Minus (Subtract)*	*Find the difference.*	$6{,}003 - 1{,}457 = 4{,}546$
$\times$	*Times (Multiply)*	*Repeated groups.*	$36 \times 24 = 864$
$\div$	*Divide*	*Split equally.*	$91 \div 7 = 13$
$=$	*Equals*	*Same value on both sides.*	$\frac{3}{6} = \frac{1}{2}$
$>$ $<$	*Greater / Less Than*	*Compares two values.*	$\frac{3}{4} > \frac{2}{5}$
$\frac{a}{b}$	*Fraction Bar*	a *parts out of* b *equal parts.*	
.	*Decimal Point*	*Separates wholes from parts.*	$0.75 = \frac{75}{100}$
$\angle$	*Angle*	*Two rays sharing a vertex.*	$\angle ABC = 45°$

Key Math Words for Grade 4

- ***Factor*** — *number you multiply by*
- ***Product*** — *answer when you multiply*
- ***Quotient*** — *answer when you divide*
- ***Remainder*** — *left over after dividing*
- ***Prime*** — *only 1 and itself as factors*
- ***Composite*** — *more than two factors*
- ***Numerator*** — *top of a fraction*
- ***Denominator*** — *bottom of a fraction*
- ***Equivalent*** — *same value, different form*
- ***Mixed Number*** — *whole + fraction*
- ***Perimeter*** — *distance around a shape*
- ***Area*** — *space inside a shape*

Multiplication Table

Knowing these facts makes every topic easier!

×	1	2	3	4	5	6	7	8	9	10	11	12
1	1	2	3	4	5	6	7	8	9	10	11	12
2	2	4	6	8	10	12	14	16	18	20	22	24
3	3	6	9	12	15	18	21	24	27	30	33	36
4	4	8	12	16	20	24	28	32	36	40	44	48
5	5	10	15	20	25	30	35	40	45	50	55	60
6	6	12	18	24	30	36	42	48	54	60	66	72
7	7	14	21	28	35	42	49	56	63	70	77	84
8	8	16	24	32	40	48	56	64	72	80	88	96
9	9	18	27	36	45	54	63	72	81	90	99	108
10	10	20	30	40	50	60	70	80	90	100	110	120
11	11	22	33	44	55	66	77	88	99	110	121	132
12	12	24	36	48	60	72	84	96	108	120	132	144

How to Use This Table

Multiply: *Find one factor in the left column, the other in the top row. Where they meet is your answer!*

Divide: *For $72 \div 8$, find 72 in the 8-row, then look up to the top row. The answer is 9!*

“*The $\times 9$ trick: multiply by 10 and subtract once. So $9 \times 7 = 70 - 7 = 63$. Works every time! And $\times 12 = \times 10 + \times 2$. Try $12 \times 6 = 60 + 12 = 72$!*”

PRACTICE & TRACK

Take Your Learning Online with ViewMath Academy!

For Parents, Teachers & Students

Love what you're reading? ViewMath Academy is your ***free online companion*** *to this book — practice what you learn, track your progress, and master every topic!*

- ***Topic Quizzes*** *— Test yourself on each topic right after you study it in this book*
- ***Interactive Lessons*** *— Revisit any concept with online lessons that match each chapter*
- ***Progress Tracking*** *— Watch your mastery grow as you work through the book*
- ***Adaptive Practice*** *— Get more questions on topics where you need extra help*
- ***Practice Tests*** *— When you're ready, take a full practice test and analyze your score online*

Scan to visit ViewMath Academy

ViewMath.com/TX-Grade4

Table of Contents

Here's what we'll explore together!

Let's learn and have fun!

CHAPTER

1

Multiply, Divide, and Patterns

What's Inside

1.1 Multiplication as a Comparison

In this lesson you will learn:

- *Understand what "times as many" means*
- *Write multiplication equations from comparison sentences*
- *Read statements like "35 is 5 times as many as 7"*

What Does "Times As Many" Mean?

*When we say **"3 times as many,"** we mean a number is multiplied by 3.*

Example:

- *Sam has 4 stickers.*
- *Mia has 3 times as many stickers as Sam.*
- *Mia has $3 \times 4 = 12$ stickers.*

Mia has 3 groups of 4. That's 12!

***Key idea:** "**Times as many**" always means **multiply**. "**More than**" means **add**. They are NOT the same!*

"If you see the words "times as many," grab your multiplication skills!"

Reading a Comparison Sentence

"24 is 6 times as many as 4."

1. *Find the two numbers being compared:* 24 *and* 4.
2. *The multiplier is* 6.
3. *Check:* $6 \times 4 = 24$ ✓

Answer: $6 \times 4 = 24$

Writing a Comparison Equation

Jake has 5 toy cars. Emma has 4 times as many toy cars as Jake. How many toy cars does Emma have?

1. *Jake has* 5 *cars (the smaller amount).*
2. *Emma has* 4 *times as many.*
3. *Multiply:* $4 \times 5 = 20$

Answer: *Emma has* 20 *toy cars.*

Multiplication as a Comparison — Practice

1. 6 times as many as 3 = ________

2. 4 times as many as 7 = ________

3. 8 times as many as 5 = ________

4. $3\times$ ________ $= 21$

5. Ben has 6 baseball cards. Luis has 4 times as many baseball cards as Ben. How many baseball cards does Luis have?

Answer: ____________ cards

6. A tree is 9 feet tall. A building is 7 times as tall as the tree. How tall is the building?

Answer: ____________ feet

Get Online
Find more at
ViewMath.com/TX-Grade4
ViewMath.com

★ 1.2 Comparison Word Problems ★

In this lesson you will learn:

- *Solve multiplicative comparison word problems*
- *Use bar models and equations with a letter for the unknown*
- *Tell the difference between additive and multiplicative comparisons*

Three Types of Comparison Problems

All multiplicative comparison problems use the equation:

big amount = multiplier × small amount

Type 1: Find the bigger amount

Sara has 8 *crayons. Tom has* 3 *times as many.*

$t = 3 \times 8 = 24$ *crayons.*

Type 2: Find the multiplier

Zoe has 7 *stickers. Maya has* 42*.*

$42 = n \times 7 \ \Rightarrow \ n = 42 \div 7 = 6$ *times as many.*

Type 3: Find the smaller amount

A book costs 4 *times as much as a pencil. The book costs* $\$12$*.*

$12 = 4 \times p \ \Rightarrow \ p = 12 \div 4 = \$3$

TIP

When you know the BIG number and the multiplier, DIVIDE to find the small number!

Finding the Bigger Amount

A pet store has 6 goldfish. It has 5 times as many guppies as goldfish. How many guppies?

1. *Goldfish:* 6 *(the smaller amount).*
2. *Guppies:* 5 *times as many.*
3. *Equation:* $g = 5 \times 6$
4. *Solve:* $g = 30$

> ***Answer:*** ***The pet store has*** 30 ***guppies.***

Finding the Smaller Amount

A book costs 4 times as much as a pencil. The book costs $\$12$. How much does the pencil cost?

1. *The book* $(\$12)$ *is* 4 *times the pencil.*
2. *Equation:* $12 = 4 \times p$
3. *Divide:* $p = 12 \div 4 = 3$

> ***Answer:*** ***The pencil costs*** $\$3$.

Get Online
Find more at
ViewMath.com/TX-Grade4
ViewMath.com

Comparison Word Problems — Practice

1. Amy has 7 stickers. Jake has 6 times as many stickers as Amy. How many stickers does Jake have?

Answer: ________________ *stickers*

2. A snake is 4 feet long. A python is 6 times as long. How long is the python?

Answer: ________________ *feet*

3. Mia has 56 beads. She has 8 times as many beads as Lily. How many beads does Lily have?

Answer: ________________ *beads*

4. A cat weighs 9 pounds. A dog weighs 36 pounds. The dog weighs how many times as much as the cat?

Answer: ________________ *times*

5. *Ryan has 5 toy cars. His sister has 3 times as many. How many toy cars do they have* ***altogether****?* __________

6. *Ella has 8 coins. After her birthday, she has 4 times as many. How many* ***new*** *coins did she get?* __________

1.3 Multi-Step Word Problems

In this lesson you will learn:

- *Solve word problems that take two or more steps*
- *Use a letter to stand for an unknown number*
- *Check if your answer makes sense using estimation*

What Is a Multi-Step Problem?

*A **multi-step word problem** needs more than one calculation to solve.*

You might need to:

- *Add **then** subtract*
- *Multiply **then** add*
- *Divide **then** subtract*
- *Or any combination of $+$, $-$, $\times$, $\div$*

***Interpreting remainders:** When you divide and get a remainder, think carefully!*

- ***"How many full groups?"** → Drop the remainder.*
- ***"How many needed?"** → Round up.*
- ***"How many left over?"** → The remainder IS the answer.*

> *"The secret to multi-step problems? Take it one step at a time! Don't try to do everything at once."*

Get Online

Find more at
ViewMath.com/TX-Grade4

ViewMath.com

Multiply, Then Add

Carlos buys 3 packs of markers. Each pack has 8 markers. He already had 5 markers at home. How many markers does Carlos have now?

1. *Markers bought:* $3 \times 8 = 24$
2. *Add markers from home:* $24 + 5 = 29$
3. *Check:* $3 \times 8 = 24$, *and* $24 + 5 = 29$ ✓

Answer: Carlos has 29 markers.

Divide, Then Subtract

A teacher has 24 pencils to share equally among 4 students. Each student gives 2 pencils to a friend. How many pencils does each student have left?

1. *Divide:* $24 \div 4 = 6$ *pencils each*
2. *Subtract:* $6 - 2 = 4$ *pencils left*

Answer: Each student has 4 pencils left.

Get Online
Find more at
ViewMath.com/TX-Grade4
ViewMath.com

Multi-Step Word Problems — Practice

1. *Amy has 5 boxes with 6 cookies each. She eats 4 cookies. How many cookies are left?*

Answer: ________________ cookies

2. *A store has 48 apples in 6 equal bags. Tom buys 2 bags. How many apples does Tom have?*

Answer: ________________ apples

3. *Sam earns $7 per hour. He works 4 hours on Saturday and 3 hours on Sunday. How much does he earn for the weekend?*

Answer: ________________ dollars

4. *There are 30 students going on a field trip. Each van holds 7 students. How many vans are needed?*

Answer: ________________ vans

5. *A baker makes 36 muffins. She puts 8 muffins in each box. She sells 3 full boxes. How many muffins are left?*

Answer: ________________ muffins

6. *Write an equation using a letter for the unknown, then solve: Maria buys 4 packs of stickers with 9 stickers each. She gives away 10 stickers. Let s = stickers left.* ______________________

★ 1.4 Factors and Factor Pairs ★

In this lesson you will learn:

- *Understand what a factor is*
- *Find all factor pairs for numbers from 1 to 100*
- *Recognize that a whole number is a multiple of each of its factors*

What Are Factors?

*A **factor** is a number that divides evenly into another number with no remainder.*

*A **factor pair** is two numbers that multiply together to make a product.*

***Example:** Find all factor pairs of 12.*

- $1 \times 12 = 12$ ✓ *So 1 and 12 are factors.*
- $2 \times 6 = 12$ ✓ *So 2 and 6 are factors.*
- $3 \times 4 = 12$ ✓ *So 3 and 4 are factors.*
- *Try 4: already found! Stop.*

***Factors of** 12: $1, 2, 3, 4, 6, 12$*

Every number has at least two factors: 1 and itself.

“When you find one factor, you automatically find its partner! If 3 is a factor of 18, then $18 \div 3 = 6$, so 6 is also a factor!”

Get Online
Find more at
ViewMath.com/TX-Grade4
ViewMath.com

Find All Factor Pairs of 24

1. $1 \times 24 = 24$ ✓
2. $2 \times 12 = 24$ ✓
3. $3 \times 8 = 24$ ✓
4. $4 \times 6 = 24$ ✓
5. *Try* 5: $24 \div 5 = 4$ *R4 — not a factor!*
6. 6 *was already found. Stop!*

Answer: Factor pairs: $(1, 24)$, $(2, 12)$, $(3, 8)$, $(4, 6)$

Find All Factors of 36

1. 1×36, 2×18, 3×12, 4×9, 6×6
2. *Try* 5: $36 \div 5 = 7$ *R1 — not a factor!*
3. 6×6: *repeated number. Stop!*

Answer: Factors of 36: $1, 2, 3, 4, 6, 9, 12, 18, 36$

Factors and Factor Pairs — Practice

1. *List all factors of 18.* ____________________

2. *List all factor pairs of 20.* ____________________

3. *List all factors of 32.* ____________________

4. *Is 7 a factor of 42?* __________

5. *Is 5 a factor of 27?* __________

6. *Maria has 28 stickers and wants to share them equally among friends with none left over. List all the ways she can divide them (the number of friends in each group).*

Answer: ____________ *friends*

★ 1.5 Multiples, Primes, and Composites ★

In this lesson you will learn:

- *Find multiples of a number*
- *Classify numbers (1–100) as prime or composite*
- *Know that 1 is neither prime nor composite*

Multiples, Primes, and Composites

Multiples

*A **multiple** of a number is the result of multiplying it by* $1, 2, 3, 4, \ldots$

Multiples of 3*:* $3, 6, 9, 12, 15, 18, \ldots$ *(they go on forever!)*

Prime Numbers

*A **prime number** has exactly* 2 *factors:* 1 *and itself.*

Primes up to 30*:* $2, 3, 5, 7, 11, 13, 17, 19, 23, 29$

Composite Numbers

*A **composite number** has **more than*** 2 *factors.*

Examples: $4, 6, 8, 9, 10, 12, 14, 15, 16, 18, 20, \ldots$

Special case: *The number* 1 *is **neither** prime nor composite (it has only* 1 *factor).*

2 *is the **only even prime number**. Every other even number is composite.*

TIP

Factors are FEWER (there's a limited list). Multiples are MANY (they go on forever)!

Is 23 Prime or Composite?

1. *Try dividing:* $23 \div 2 = 11$ *R1,* $23 \div 3 = 7$ *R2,* $23 \div 4 = 5$ *R3*
2. *No whole number (other than* 1 *and* 23*) divides evenly.*
3. *Exactly* 2 *factors:* 1 *and* 23*.*

> ✔ ***Answer:*** 23 ***is prime.***

Is 15 Prime or Composite?

1. $15 \div 3 = 5$ ✓ *Found a factor besides* 1 *and* 15*!*
2. *Factors of* 15*:* $1, 3, 5, 15$ *— that's* 4 *factors.*
3. *More than* 2 *factors.*

> ✔ ***Answer:*** 15 ***is composite.***

Multiples, Primes, and Composites — Practice

1. List the first 5 multiples of 6. ____________________

2. Is 14 a multiple of 3? __________

3. Is 17 prime or composite? __________

4. Is 27 prime or composite? __________

5. The number 1 is a prime number. True ◯ False ◯

6. Name all the prime numbers between 10 and 20. ____________________

1.6 Number and Shape Patterns

In this lesson you will learn:

- *Generate a number pattern that follows a given rule*
- *Identify features of a pattern not stated in the rule*
- *Recognize and extend shape patterns*

Number Patterns and Rules

*A **pattern** is a list of numbers (or shapes) that follows a **rule**.*

Common rules:

- *"Add 5"* $\rightarrow 3, 8, 13, 18, 23, \ldots$
- *"Subtract 2"* $\rightarrow 20, 18, 16, 14, 12, \ldots$
- *"Multiply by 2"* $\rightarrow 1, 2, 4, 8, 16, \ldots$

Hidden features: *A pattern can have features not stated in the rule.*

Rule: "Start at 1. Add 2." $\rightarrow 1, 3, 5, 7, 9, 11, \ldots$

*Hidden feature: All the numbers are **odd**! The rule doesn't say that, but it's true.*

> *"After you find the rule, always look for hidden features! Are all the numbers even? Odd? Do they alternate?"*

Get Online

Find more at
ViewMath.com/TX-Grade4

ViewMath.com

Find the Rule and Continue

Pattern: $4,\ 9,\ 14,\ 19,\ 24,\ \ldots$

1. *Find the differences:* $9-4=5,\quad 14-9=5,\quad 19-14=5$
2. *Rule:* **Add** 5
3. *Next two:* $24+5=29,\quad 29+5=34$

Answer: Rule: Add 5. **Next two numbers:** $29,\quad 34$

Spot the Hidden Feature

Rule: Start at 2. **Add** 4 **each time.**

Pattern: $2,\ 6,\ 10,\ 14,\ 18,\ 22,\ \ldots$

Solution:

*All the numbers are **even**. The rule says "add 4," but it doesn't mention that every number is even — that's a hidden feature!*

Number and Shape Patterns — Practice

1. *Rule: Start at* 3. *Add* 4. *Write the first* 6 *numbers.* ______________________

2. *Find the rule:* 5, 10, 15, 20, 25, ... _____________

3. *Continue the pattern:* 64, 32, 16, 8, ______, ______

4. *Rule: Start at* 1. *Add* 3. *What hidden feature do you notice?* ______________________

5. *Fill in the missing numbers:* 7, 14, 21, ______, ______, 42

6. *Rule: Start at* 100. *Subtract* 9. *Write the first* 5 *numbers.* ______________________

Get Online
Find more at
ViewMath.com/TX-Grade4
ViewMath.com

CHAPTER

2

Big Numbers and Place Value

What's Inside

★ 2.1 Place Value to One Million ★

In this lesson you will learn:

- *Name each place value from ones to hundred-thousands*
- *Extend place value understanding to **millions** and **billions***
- *Tell the value of any digit based on its position*

The Place Value Chart — Extended to Billions

*Every digit in a number has a **place** and a **value**. Numbers are grouped into **periods** of three digits.*

Billions Period			Millions Period			Thousands Period		
H-B	T-B	B	H-M	T-M	M	H-Th	T-Th	Th

Example: *In* **5,204,738,000***:*

- 5 *is in the **billions** place → value is* 5,000,000,000.
- 2 *is in the **hundred-millions** place → value is* 200,000,000.
- 7 *is in the **hundred-thousands** place → value is* 700,000.

*A **period** is a group of three digits separated by commas: ones, thousands, millions, billions.*

"Think of place value like an address — the POSITION of a digit tells you its VALUE. The digit 3 can mean 3, 300, or even 3,000,000,000!"

Get Online

Find more at
ViewMath.com/TX-Grade4

ViewMath.com

Reading a Number in the Millions

Write $47{,}063{,}200$ ***in word form.***

1. *Millions period:* $47 \rightarrow$ *"forty-seven million"*
2. *Thousands period:* $063 \rightarrow$ *"sixty-three thousand"*
3. *Ones period:* $200 \rightarrow$ *"two hundred"*

Answer: Forty-seven million, sixty-three thousand, two hundred

Finding a Digit's Value in the Billions

In $6{,}381{,}000{,}000$***, what is the value of the digit*** 3***?***

1. *Find* 3 *in the number:* $6{,}\mathbf{3}81{,}000{,}000$
2. *It is in the* ***hundred-millions*** *place.*
3. *Value:* $3 \times 100{,}000{,}000 = 300{,}000{,}000$

Answer: The value of the digit 3 ***is*** $300{,}000{,}000$***.***

Place Value to Billions — Practice

1. In 463,825, what place is the digit 6 in?

______

2. What is the value of the 7 in 72,405,000?

______

3. What is the value of the 9 in 9,000,000,000?

______

4. Write 605,090 in word form. ______

5. Write "three billion, two hundred million" in standard form. ______

6. The world population is about 8,100,000,000. What is the value of the digit 8 in this number?

Answer: ______

★ 2.2 Expanded Form ★

In this lesson you will learn:

- *Write numbers in expanded form showing the value of each digit*
- *Switch between standard form and expanded form*
- *Handle zeros correctly in expanded form*

What Is Expanded Form?

Expanded form *breaks a number into the value of each digit, added together.*

Example: $4{,}725$

$4{,}000$	$+$	700	$+$	20	$+$	5
Thousands		*Hundreds*		*Tens*		*Ones*

So $4{,}725 = 4{,}000 + 700 + 20 + 5$.

What about zeros? *When a digit is* 0*, skip it! Its value is* 0.

Example: $70{,}502 = 70{,}000 + 500 + 2$

" Think of expanded form like unpacking a suitcase — you take every piece out so you can see each digit's true value! "

Writing in Expanded Form

Write** $425{,}318$ **in expanded form.

1. 4 *hundred-thousands* $= 400{,}000$
2. 2 *ten-thousands* $= 20{,}000$
3. 5 *thousands* $= 5{,}000$
4. 3 *hundreds* $= 300$, 1 *ten* $= 10$, 8 *ones* $= 8$

Answer: $425{,}318 = 400{,}000 + 20{,}000 + 5{,}000 + 300 + 10 + 8$

Going from Expanded to Standard Form

Write** $600{,}000 + 30{,}000 + 200 + 50 + 9$ **in standard form.

1. *Add from left to right:* $600{,}000 + 30{,}000 = 630{,}000$
2. $630{,}000 + 200 + 50 + 9 = 630{,}259$

Answer: $630{,}259$

Get Online
Find more at
ViewMath.com/TX-Grade4
ViewMath.com

Expanded Form — Practice

1. Write $6{,}231$ in expanded form. ________________________
2. Write $340{,}915$ in expanded form. ________________________
3. Write $802{,}060$ in expanded form. ________________________
4. Write $500{,}000 + 9{,}000 + 300 + 20$ in standard form. ______________
5. Write $400{,}000 + 70{,}000 + 8$ in standard form. ______________
6. $58{,}407 = 50{,}000+$ _______ $+400 + 7$. What is the missing value?

Get Online
Find more at
ViewMath.com/TX-Grade4
ViewMath.com

2.3 Compare Big Numbers

In this lesson you will learn:

- *Compare two multi-digit numbers using $>$, $=$, and $<$*
- *Compare numbers in the **millions and billions***
- *Order numbers from least to greatest*

Comparing Numbers — Even Really Big Ones!

*To compare two numbers, look at each place value **from LEFT to RIGHT**.*

Step 1: *If the numbers have a different number of digits, the one with **more digits** is bigger.*

Step 2: *If same number of digits, compare **left to right** until a digit differs.*

Example: *Compare* $2{,}450{,}000{,}000$ *and* $2{,}380{,}000{,}000$.

- *Billions: both* 2 *— same! Move right.*
- *Hundred-millions:* 4 *vs.* 3 *—* $4 > 3$. ***Stop!***

Answer: $2{,}450{,}000{,}000 \quad > \quad 2{,}380{,}000{,}000$

Ordering: *To put numbers in order, compare them two at a time.*

TIP

*The comparison symbol is like an alligator mouth — it always opens toward the **bigger** number!*

Comparing Millions

Compare $47{,}500{,}000$ ***and*** $47{,}350{,}000$.

1. *Both have* 8 *digits. Compare left to right.*
2. *T-M:* $4 = 4$. *Millions:* $7 = 7$.
3. *Hundred-thousands:* 5 *vs.* $3 \to 5 > 3$. *Stop!*

✔ ***Answer:*** $47{,}500{,}000 \quad > \quad 47{,}350{,}000$

Ordering Billions

Order from least to greatest: $8{,}100{,}000{,}000 \quad 850{,}000{,}000 \quad 8{,}500{,}000{,}000$

1. $850{,}000{,}000$ *has* 9 *digits; the others have* 10 *digits. It's the smallest.*
2. *Compare* $8{,}100{,}000{,}000$ *and* $8{,}500{,}000{,}000$*: hundred-millions* $1 < 5$.

✔ ***Answer:*** $850{,}000{,}000 < 8{,}100{,}000{,}000 < 8{,}500{,}000{,}000$

Get Online
Find more at
ViewMath.com/TX-Grade4
ViewMath.com

Compare Big Numbers — Practice

1. 67,432 ________ 67,234

2. 5,410,000 ________ 5,420,000

3. 3,000,000,000 ________ 2,999,999,999

4. 83,450,000 ________ 83,450,000

5. *Order from least to greatest:* 750,000 7,500,000 75,000 7,500 ____________________________

6. *Texas has about* 30,000,000 *people and California has about* 39,000,000 *people. Which state has more people?*

Answer: ________________ *State*

★ 2.4 Ten Times Bigger, Ten Times Smaller ★

In this lesson you will learn:

- *Understand that a digit one place to the left is 10 times bigger*
- *Use multiplication by 10 and division by 10 to show place value relationships*
- *Explain how the same digit has different values in different places*

The $\times 10$ Relationship

When a digit moves ***one place to the LEFT****, its value becomes 10* ***times bigger****. When a digit moves* ***one place to the RIGHT****, its value becomes 10* ***times smaller****.*

TTh	Th	H	T	O
				5
			5	0
		5	0	0
	5	0	0	0
5	0	0	0	0

- $5 \times 10 = 50$ *(ones → tens)*
- $50 \times 10 = 500$ *(tens → hundreds)*
- $500 \times 10 = 5{,}000$ *(hundreds → thousands)*
- $5{,}000 \times 10 = 50{,}000$ *(thousands → ten thousands)*

Quick rule: *Multiply by 10 = add a zero. Divide by 10 = remove a zero.*

> *"Every time a digit slides one place LEFT, it gets 10 times more powerful — like leveling up!"*

How Many Times Bigger?

In 3,344, the digit 3 appears twice. How does the value of the 3 in the thousands place compare to the 3 in the hundreds place?

1. 3 in the thousands place $= 3{,}000$
2. 3 in the hundreds place $= 300$
3. $3{,}000 \div 300 = 10$

Answer: ***The 3 in the thousands place is 10 times the value of the 3 in the hundreds place.***

Fill in the Blank

$700 = \_\_\_\_ \times 70$

1. 700 is one place to the left of 70.
2. $70 \times 10 = 700$ ✓

Answer: $700 = 10 \times 70$

Get Online
Find more at
ViewMath.com/TX-Grade4
ViewMath.com

Ten Times Bigger, Ten Times Smaller — Practice

1. $34 \times 10 =$ ________

2. $560 \times 10 =$ ________

3. $4{,}200 \div 10 =$ ________

4. $90{,}000 \div 10 =$ ________

5. *In* $6{,}662$, *how many times greater is the value of the* 6 *in the thousands place than the* 6 *in the tens place?*

____

6. *Complete:* $800 =$ ________ $\times 80$

★ 2.5 Rounding Whole Numbers ★

In this lesson you will learn:

- *Round multi-digit numbers to any place through* ***millions and billions***
- *Know when to round up and when to round down*
- *Use rounding to estimate and check answers*

How Rounding Works — Even with Really Big Numbers

Rounding *replaces a number with a nearby number that ends in zeros.*

Steps:

1. ***Underline*** *the digit in the place you are rounding to.*
2. *Look at the digit* ***one place to the RIGHT****.*
3. *If it is* **0, 1, 2, 3, 4** $\rightarrow$ ***round DOWN*** *(keep the digit).*
4. *If it is* **5, 6, 7, 8, 9** $\rightarrow$ ***round UP*** *(add 1).*
5. *Change all digits to the right to zeros.*

This works for ANY place *— thousands, hundred-thousands, millions, or even billions!*

Example: *Round 4,738,000 to the nearest million.*

Millions digit = 4. Look right: hundred-thousands digit = 7.

$7 \geq 5$*, so round UP* $\rightarrow$ **5,000,000**.

"Here's my rounding rhyme: "5 or more, let it soar! 4 or less, let it rest!" It works for ALL place values!"

Get Online

Find more at ViewMath.com/TX-Grade4

Round to the Nearest Hundred-Thousand

Round $682{,}419$ **to the nearest hundred-thousand.**

1. *Hundred-thousands digit* $= 6$
2. *Look right: ten-thousands digit* $= 8$
3. $8 \geq 5$*, so round UP:* 6 *becomes* 7
4. *Replace everything to the right with zeros*

Answer: $682{,}419$ **rounds to 700,000**

Round to the Nearest Billion

Round $3{,}482{,}000{,}000$ **to the nearest billion.**

1. *Billions digit* $= 3$
2. *Look right: hundred-millions digit* $= 4$
3. $4 < 5$*, so round DOWN:* 3 *stays as* 3
4. *Replace everything to the right with zeros*

Answer: $3{,}482{,}000{,}000$ **rounds to 3,000,000,000**

Rounding Whole Numbers — Practice

1. *Round* 347 *to the nearest hundred.* __________

2. *Round* 67,812 *to the nearest thousand.* __________

3. *Round* 482,500 *to the nearest hundred-thousand.* __________

4. *Round* 7,340,000 *to the nearest million.* __________

5. *Round* 5,600,000,000 *to the nearest billion.* __________

6. *A state has about* 29,530,000 *people. The governor says "about* 30 *million live here." To what place did the governor round?*

Answer: ______________

CHAPTER

3

Add, Subtract, Multiply & Divide

What's Inside

★ 3.1 Add Whole Numbers ★

In this lesson you will learn:

- Add multi-digit whole numbers using the standard algorithm
- Regroup (carry) when a column adds to 10 or more
- Add numbers up to the thousands place

The Standard Addition Algorithm

To add multi-digit numbers, **line up the digits by place value** and add column by column from right to left.

$$\begin{array}{r} {}^{1}\ {}^{1}\ \ \\ 3\ 4\ 7 \\ +\ 2\ 8\ 5 \\ \hline 6\ 3\ 2 \end{array}$$

Regrouping (carrying): When a column adds to 10 or more, write the ones digit and **carry** the tens digit to the next column.

- Ones: $7 + 5 = 12$. Write 2, carry 1.
- Tens: $4 + 8 + 1 = 13$. Write 3, carry 1.
- Hundreds: $3 + 2 + 1 = 6$.

“ Always start from the **right** (ones column) and work left. Don't forget to add any carried digits! ”

Add Without Regrouping: $432 + 251$

1. *Ones:* $2 + 1 = 3$
2. *Tens:* $3 + 5 = 8$
3. *Hundreds:* $4 + 2 = 6$

Answer: $432 \quad + \quad 251 \quad = \quad 683$

Add Four-Digit Numbers: $3{,}748 + 2{,}695$

1. *Ones:* $8 + 5 = 13$. *Write* 3, *carry* 1.
2. *Tens:* $4 + 9 + 1 = 14$. *Write* 4, *carry* 1.
3. *Hundreds:* $7 + 6 + 1 = 14$. *Write* 4, *carry* 1.
4. *Thousands:* $3 + 2 + 1 = 6$.

Answer: $3{,}748 \quad + \quad 2{,}695 \quad = \quad 6{,}443$

Add Whole Numbers Practice

1. $524 + 361 =$ __________

2. $739 + 148 =$ __________

3. $2{,}456 + 3{,}387 =$ __________

4. $4{,}809 + 1{,}763 =$ __________

5. *A school collected* 1,258 *cans in January and* 1,475 *cans in February. How many cans were collected in all?*

____

6. *The library has* 3,692 *fiction books and* 2,548 *non-fiction books. How many books does the library have in total?* __________

Get Online
Find more at
ViewMath.com/TX-Grade4
ViewMath.com

★ 3.2 Subtract Whole Numbers ★

In this lesson you will learn:

- *Subtract multi-digit whole numbers using the standard algorithm*
- *Regroup (borrow) when the top digit is smaller than the bottom digit*
- *Subtract across zeros*

The Standard Subtraction Algorithm

To subtract, ***line up digits by place value*** *and subtract column by column from right to left.*

$$\begin{array}{r} 5\ 3\ 4 \\ -\ 2\ 7\ 8 \\ \hline 2\ 5\ 6 \end{array}$$

Regrouping (borrowing): *When the top digit is smaller than the bottom digit,* ***borrow*** *1 from the next column to the left. That adds 10 to the current column.*

- *Ones:* $4 - 8$*? Borrow!* $14 - 8 = 6$
- *Tens:* $2 - 7$*? Borrow!* $12 - 7 = 5$
- *Hundreds:* $4 - 2 = 2$

> TIP
> *Check your answer with addition:* $256 + 278 = 534$ ✓

Subtract Without Regrouping: $685 - 243$

1. *Ones:* $5 - 3 = 2$
2. *Tens:* $8 - 4 = 4$
3. *Hundreds:* $6 - 2 = 4$

Answer: $685 - 243 = 442$

Subtract Across Zeros: $4{,}003 - 1{,}867$

When zero is in the top number, keep borrowing left until you find a digit you can borrow from.

1. *Ones:* $3 - 7$*? Borrow from tens — but tens is* 0*! Go to hundreds — also* 0*! Go to thousands.*
2. *Borrow from* 4 *thousands* $\rightarrow$ 3 *thousands. Work back:* 0 *hundreds becomes* 10*, then borrow to get* 9 *hundreds and* 10 *tens, then borrow to get* 9 *tens and* 13 *ones.*
3. *Ones:* $13 - 7 = 6$*, Tens:* $9 - 6 = 3$*, Hundreds:* $9 - 8 = 1$*, Thousands:* $3 - 1 = 2$

Answer: $4{,}003 - 1{,}867 = 2{,}136$

> *After subtracting, always check by adding: answer + bottom number should equal the top number!*

Get Online
Find more at
ViewMath.com/TX-Grade4
ViewMath.com

Subtract Whole Numbers Practice

1. $876 - 432 =$ __________

2. $615 - 289 =$ __________

3. $5{,}274 - 1{,}836 =$ __________

4. $7{,}005 - 2{,}468 =$ __________

5. *A farmer had* 3,500 *apples. He sold* 1,275 *at the market. How many apples does he have left?* __________

6. *The town library had* 8,042 *visitors last year and* 5,687 *visitors this year. How many fewer visitors came this year?* __________

★ 3.3 Multiply by a One-Digit Number ★

In this lesson you will learn:

- *Multiply a two-, three-, or four-digit number by a one-digit number*
- *Use partial products and the standard algorithm*
- *Regroup (carry) when a product is 10 or more*

Multiply Using Place Value

Break the bigger number into place values, multiply each part by the one-digit factor, and add the ***partial products***.

Example: 5×347

Break 347 *into* $300 + 40 + 7$:

	300	40	7
5	5×300 $= 1{,}500$	5×40 $= 200$	5×7 $= 35$

$1{,}500 + 200 + 35 =$ **1,735**

Standard algorithm: *Multiply digit by digit from right to left, carrying when needed — just like addition, but you multiply first.*

“The area model shows WHY multiplication works. The standard algorithm is the FAST way to do it!”

Standard Algorithm: 7×463

1. *Ones:* $7 \times 3 = 21$. *Write* 1*, carry* 2.
2. *Tens:* $7 \times 6 = 42$*, plus* $2 = 44$. *Write* 4*, carry* 4.
3. *Hundreds:* $7 \times 4 = 28$*, plus* $4 = 32$. *Write* 32.

Answer: $7 \times 463 = 3{,}241$

Four-Digit Multiplication: $4 \times 2{,}136$

1. *Ones:* $4 \times 6 = 24$. *Write* 4*, carry* 2.
2. *Tens:* $4 \times 3 = 12$*, plus* $2 = 14$. *Write* 4*, carry* 1.
3. *Hundreds:* $4 \times 1 = 4$*, plus* $1 = 5$. *Write* 5.
4. *Thousands:* $4 \times 2 = 8$. *Write* 8.

Answer: $4 \times 2{,}136 = 8{,}544$

Multiply by One Digit Practice

1. $6 \times 83 =$ ________

2. $5 \times 247 =$ ________

3. $8 \times 619 =$ ________

4. $3 \times 4{,}215 =$ ________

5. *A bookstore ordered* 7 *boxes with* 384 *books in each box. How many books were ordered?* ________

6. *Each floor of a building has* 1,250 *square feet. The building has* 6 *floors. What is the total square footage?*

____

★ 3.4 Multiply Two-Digit by Two-Digit ★

◎ In this lesson you will learn:

- *Multiply a two-digit number by another two-digit number*
- *Use partial products and the standard algorithm*
- *Understand the role of the placeholder zero when multiplying by tens*

Two-Digit × Two-Digit with Partial Products

Break ***both*** *numbers into tens and ones, then multiply each pair.*

Example: 23×15

Break into: $(20 + 3) \times (10 + 5)$

	10	5
20	20×10 $= 200$	20×5 $= 100$
3	3×10 $= 30$	3×5 $= 15$

Add all four parts: $200 + 100 + 30 + 15 =$ **345**

Get Online
Find more at
ViewMath.com/TX-Grade4
ViewMath.com

> *"The standard algorithm does the same thing in two rounds: first multiply by the ones digit, then by the tens digit, and add!"*

Partial Products: 34×27

Break apart: $(30 + 4) \times (20 + 7)$

1. $30 \times 20 = 600$
2. $30 \times 7 = 210$
3. $4 \times 20 = 80$
4. $4 \times 7 = 28$
5. *Add:* $600 + 210 + 80 + 28 = 918$

Answer: $34 \times 27 = 918$

Standard Algorithm: 56×43

1. ***Round 1 — Multiply by** 3 **(ones):***

$3 \times 6 = 18$, *write* 8, *carry* 1.

$3 \times 5 = 15 + 1 = 16$.

First row: 168

2. ***Round 2 — Multiply by** 40 **(tens):***

Write 0 *in the ones place.*

$4 \times 6 = 24$, *write* 4, *carry* 2.

$4 \times 5 = 20 + 2 = 22$.

Second row: $2{,}240$

3. *Add the rows:* $168 + 2{,}240 = 2{,}408$

Get Online
Find more at
ViewMath.com/TX-Grade4
ViewMath.com

Answer: $56 \times 43 = 2{,}408$

Two-Digit Multiplication Practice

1. $21 \times 14 =$ __________
2. $35 \times 26 =$ __________
3. $48 \times 53 =$ __________
4. $67 \times 39 =$ __________

5. *A classroom has* 24 *desks. Each desk has* 18 *pencils in the pencil box. How many pencils are there in all?*

____

6. *Tickets to the school play cost* \$12 *each. If* 75 *people attend, how much money is collected?* __________

3.5 Divide with Remainders

In this lesson you will learn:

- *Divide up to a four-digit number by a one-digit divisor*
- *Understand what a remainder is and use long division*
- *Interpret remainders in word problems*

Long Division: Divide, Multiply, Subtract, Bring Down

To divide, use these four steps and repeat:

1. ***Divide:*** *How many times does the divisor go into the current digits?*
2. ***Multiply:*** *Divisor × that quotient digit.*
3. ***Subtract:*** *Find what's left.*
4. ***Bring down:*** *Bring down the next digit and repeat.*

$$\underbrace{25}_{\text{dividend}} \div \underbrace{4}_{\text{divisor}} = \underbrace{6}_{\text{quotient}}\ R\ \underbrace{1}_{\text{remainder}}$$

Check: *(Quotient × Divisor) + Remainder = Dividend*

$(6 \times 4) + 1 = 25$ ✓

TIP

*Remember the steps: **D**oes **M**cDonald's **S**ell **B**urgers? — **D**ivide, **M**ultiply, **S**ubtract, **B**ring down!*

Get Online
Find more at
ViewMath.com/TX-Grade4
ViewMath.com

Long Division: $347 \div 5$

1. $3 \div 5$? *Can't! Look at* 34.
2. $34 \div 5 = 6$ *(since* $5 \times 6 = 30$*). Subtract:* $34 - 30 = 4$.
3. *Bring down the* 7*: now have* 47.
4. $47 \div 5 = 9$ *(since* $5 \times 9 = 45$*). Subtract:* $47 - 45 = 2$.
5. *No more digits. Remainder is* 2.

Answer: $347 \div 5 = 69\ R2$

Check: $(69 \times 5) + 2 = 345 + 2 = 347$ ✓

Four-Digit Division: $2{,}538 \div 6$

1. $2 \div 6$? *Can't. Look at* 25.
2. $25 \div 6 = 4$ $(6 \times 4 = 24)$. *Subtract:* $25 - 24 = 1$.
3. *Bring down* 3*: have* 13. $13 \div 6 = 2$ $(6 \times 2 = 12)$. *Subtract:* 1.
4. *Bring down* 8*: have* 18. $18 \div 6 = 3$ $(6 \times 3 = 18)$. *Subtract:* 0.

Answer: $2{,}538 \div 6 = 423$

Check: $423 \times 6 = 2{,}538$ ✓

> *"The remainder is ALWAYS less than the divisor. If it's not, your quotient digit is too small!"*

Division Practice

1. $96 \div 4 =$ __________

2. $175 \div 3 =$ __________

3. $648 \div 8 =$ __________

4. $1{,}539 \div 7 =$ __________

5. *You have* 250 *stickers to put equally into* 6 *bags. How many stickers go in each bag, and how many are left over?* __________

6. *A school bus holds* 45 *students. If* 200 *students need a ride, how many buses are needed?* __________

Get Online
Find more at
ViewMath.com/TX-Grade4
ViewMath.com

3.6 Estimate to Check Answers

In this lesson you will learn:

- *Use rounding and compatible numbers to estimate answers*
- *Check the reasonableness of exact answers using estimation*
- *Apply estimation to all four operations*

Estimate by Rounding

*To **estimate**, round each number to make the math easy, then compute.*

Addition: $587 + 324 \approx 600 + 300 = 900$ *(exact:* 911 ✓*)*

Subtraction: $8{,}243 - 3{,}891 \approx 8{,}000 - 4{,}000 = 4{,}000$ *(exact:* $4{,}352$ ✓*)*

Multiplication: $68 \times 42 \approx 70 \times 40 = 2{,}800$ *(exact:* $2{,}856$ ✓*)*

Compatible numbers for division: *Pick nearby numbers that divide evenly.*

$317 \div 4 \approx 320 \div 4 = 80$ *(exact:* 79 *R1* ✓*)*

*If your exact answer is close to the estimate, it is **reasonable**. If not, check your work!*

“Estimation is like a math superpower — it catches mistakes FAST, even before you solve the whole problem!”

Estimate a Product: 38×52

1. *Round:* $38 \approx 40$ *and* $52 \approx 50$
2. *Estimate:* $40 \times 50 = 2{,}000$
3. *Exact answer:* $38 \times 52 = 1{,}976$
4. *Compare:* $1{,}976$ *is close to* $2{,}000$ ✓

Answer: Estimate: about $2{,}000$. **The answer** $1{,}976$ **is reasonable.**

Estimate a Quotient: $458 \div 9$

1. *Find a compatible number near* 458 *that divides easily by* 9.
2. $450 \div 9 = 50$ — *easy!*
3. *Exact answer:* $458 \div 9 = 50$ *R8.*
4. *Compare: The estimate matches!* ✓

Answer: Estimate: about 50.

Estimation Practice

1. *Estimate* $493 + 318$ *by rounding to the nearest hundred.* ______________

2. *Estimate* $7{,}215 - 2{,}893$ *by rounding to the nearest thousand.* ______________

3. *Estimate* 47×63 *by rounding to the nearest ten.* ______________

4. *Use compatible numbers to estimate* $247 \div 5$. ______________

5. *A student calculated* $236 + 589 = 1{,}125$. *Estimate to check: is this reasonable?* ______________

6. *A bakery makes* 28 *trays of muffins with* 36 *muffins on each tray. Estimate the total number of muffins.*

____

CHAPTER

4

Fractions and Decimals

What's Inside

★ 4.1 Equal Fractions ★

In this lesson you will learn:

- Understand what equivalent fractions are
- Multiply or divide both parts by the same number to make equal fractions
- Simplify fractions to their simplest form

The Golden Rule of Equal Fractions

Equivalent fractions *are fractions that look different but show the* ***same amount.***

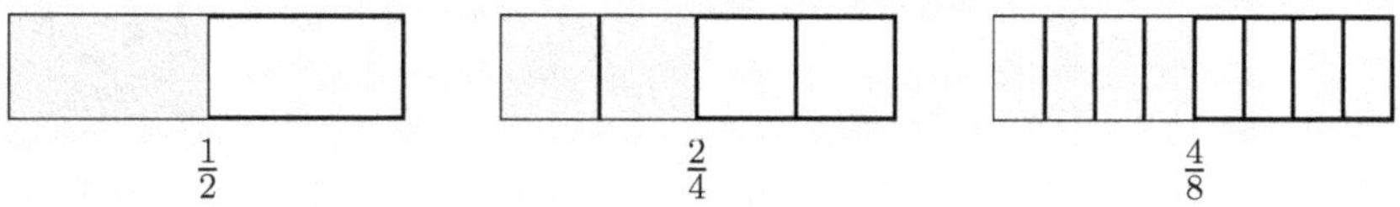

All three show the same amount shaded, so $\frac{1}{2} = \frac{2}{4} = \frac{4}{8}$.

To make an equivalent fraction: *Multiply (or divide) the* ***numerator*** *AND the* ***denominator*** *by the* ***same number.***

$$\frac{1}{3} = \frac{1 \times 4}{3 \times 4} = \frac{4}{12} \qquad \frac{6}{8} = \frac{6 \div 2}{8 \div 2} = \frac{3}{4}$$

Whatever you do to the top, you MUST do to the bottom!

“Cutting a pizza into more slices doesn't give you more pizza — the pieces are just smaller!”

Find an Equivalent Fraction for $\frac{2}{5}$

1. *Pick a number to multiply by. Let's use* 3.
2. *Multiply the numerator:* $2 \times 3 = 6$.
3. *Multiply the denominator:* $5 \times 3 = 15$.

Answer: $\frac{2}{5} = \frac{6}{15}$

Simplify $\frac{8}{12}$

1. *Find a number that divides evenly into both* 8 *and* 12*. Try* 4.
2. *Divide the numerator:* $8 \div 4 = 2$.
3. *Divide the denominator:* $12 \div 4 = 3$.

Answer: $\frac{8}{12} = \frac{2}{3}$

Get Online
Find more at
ViewMath.com/TX-Grade4
ViewMath.com

Equal Fractions — Practice

1. $\frac{1}{4} = \frac{?}{8}$ *? =* ________

2. $\frac{3}{5} = \frac{?}{15}$ *? =* ________

3. $\frac{6}{10} = \frac{?}{5}$ *? =* ________

4. *Simplify* $\frac{4}{12} =$ ________

5. *Write two fractions equivalent to* $\frac{2}{3}$. ________________

6. A recipe calls for $\frac{3}{4}$ *cup of flour. Maya wants to use a measuring cup marked in eighths. How many eighths does she need?*

Answer: ________________ *eighths*

4.2 Compare Fractions

In this lesson you will learn:

- *Compare fractions using common denominators and benchmarks*
- *Use >, <, and = symbols to compare fractions*
- *Order fractions from least to greatest*

Four Strategies for Comparing Fractions

Strategy 1 — Same Denominator: *Compare the numerators directly.*

$$\frac{3}{8} < \frac{5}{8} \quad \textit{because } 3 < 5$$

$\frac{3}{8}$

$\frac{5}{8}$

Strategy 2 — Same Numerator: *The fraction with the* ***smaller denominator*** *is* ***bigger*** *(bigger pieces!).*

$$\frac{2}{3} > \frac{2}{6} \quad \textit{because thirds are bigger than sixths}$$

Strategy 3 — Benchmark $\frac{1}{2}$: *Check if each fraction is more or less than $\frac{1}{2}$.*

- $\frac{3}{8}$ *is less than* $\frac{1}{2}$ *but* $\frac{5}{6}$ *is greater than* $\frac{1}{2}$
- *So* $\frac{3}{8} < \frac{5}{6}$

Strategy 4 — Common Denominators: *Rewrite both fractions with the same denominator, then compare numerators.*

Get Online
Find more at
ViewMath.com/TX-Grade4
ViewMath.com

Compare $\frac{5}{6}$ and $\frac{7}{8}$ Using Common Denominators

1. *Find a common denominator:* $6 \times 8 = 48$ *(or LCD* $= 24$*).*
2. *Convert:* $\frac{5}{6} = \frac{5 \times 4}{6 \times 4} = \frac{20}{24}$
3. *Convert:* $\frac{7}{8} = \frac{7 \times 3}{8 \times 3} = \frac{21}{24}$
4. *Compare numerators:* $20 < 21$

Answer: $\frac{5}{6} \quad < \quad \frac{7}{8}$

TIP

To check if a fraction is more than $\frac{1}{2}$, double the numerator. If the result is more than the denominator, the fraction is more than $\frac{1}{2}$.

Compare Fractions — Practice

1. $\frac{3}{5}$ ______ $\frac{4}{5}$ (>, <, or =)

2. $\frac{2}{3}$ ______ $\frac{2}{6}$ (>, <, or =)

3. $\frac{4}{8}$ ______ $\frac{1}{2}$ (>, <, or =)

4. $\frac{3}{4}$ ______ $\frac{5}{6}$ (>, <, or =)

5. Order from least to greatest: $\frac{1}{2}$, $\frac{3}{8}$, $\frac{3}{4}$ ____________________

6. Lily ate $\frac{2}{5}$ of a pie and Noah ate $\frac{3}{10}$ of the same pie. Who ate more?

Answer: ______________ name

Get Online
Find more at
ViewMath.com/TX-Grade4
ViewMath.com

★ 4.3 Build and Break Apart Fractions ★

In this lesson you will learn:

- *Write any fraction as a sum of unit fractions*
- *Decompose a fraction into different sums with the same denominator*
- *Compose fractions by adding parts together*

Every Fraction Is a Sum of Unit Fractions

*A **unit fraction** has 1 as the numerator: $\frac{1}{2}, \frac{1}{3}, \frac{1}{4}, \frac{1}{5}, \ldots$*

***Any fraction** $\frac{a}{b}$ is made of a copies of $\frac{1}{b}$:*

$$\frac{3}{4} = \frac{1}{4} + \frac{1}{4} + \frac{1}{4}$$

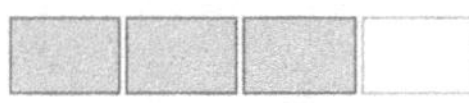

*You can also **break apart** a fraction in different ways (the denominator stays the same):*

$$\frac{5}{6} = \frac{2}{6} + \frac{3}{6} \quad \text{or} \quad \frac{5}{6} = \frac{1}{6} + \frac{4}{6}$$

When you build or break apart, the denominator NEVER changes!

“Think of pieces of the same pie. You can group slices any way you like — the total stays the same!”

Get Online

Find more at
ViewMath.com/TX-Grade4

Decompose $\frac{7}{8}$ in Two Different Ways

1. Way 1: Split into 3 and 4. $\frac{7}{8} = \frac{3}{8} + \frac{4}{8}$
2. Way 2: Split into 5 and 2. $\frac{7}{8} = \frac{5}{8} + \frac{2}{8}$
3. Check: numerators add to 7 in each. ✓

Answer: $\frac{7}{8} = \frac{3}{8} + \frac{4}{8} = \frac{5}{8} + \frac{2}{8}$

Build and Break Apart Fractions – Practice

1. Write $\frac{4}{5}$ as a sum of unit fractions. ______________

2. Decompose $\frac{5}{8}$ in two different ways.

Way 1: ______________ Way 2: ______________

3. $\frac{2}{6} + \frac{3}{6} =$ ______

4. $\frac{1}{4} + \frac{1}{4} + \frac{1}{4} =$ ______

5. True or False: $\frac{3}{5} + \frac{3}{5} = \frac{6}{10}$ ______

6. A jar is $\frac{6}{8}$ full of marbles. Sam takes out $\frac{2}{8}$. What fraction of the jar still has marbles?

Answer: ______________ fraction

Get Online
Find more at
ViewMath.com/TX-Grade4
ViewMath.com

4.4 Add and Subtract Fractions

In this lesson you will learn:

- *Add fractions with the same denominator*
- *Subtract fractions with the same denominator*
- *Simplify answers when possible*

Adding and Subtracting with Like Denominators

When fractions have the ***same denominator****, add or subtract the* ***numerators*** *and keep the denominator the same.*

Adding:

$$\frac{2}{8}+\frac{3}{8}=\frac{2+3}{8}=\frac{5}{8}$$

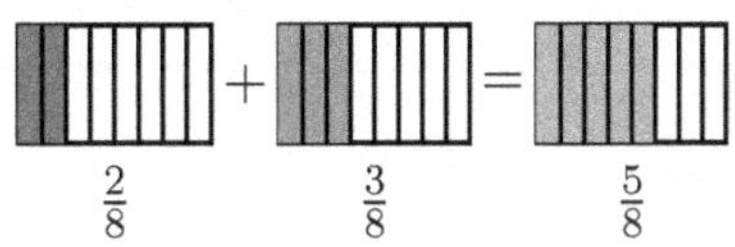

Subtracting:

$$\frac{7}{10}-\frac{3}{10}=\frac{7-3}{10}=\frac{4}{10}=\frac{2}{5}$$

Why does the denominator stay the same? *The denominator is the* ***name of the pieces****. If you have 2 eighths and add 3 eighths, you get 5 eighths — the pieces don't change size!*

Add or subtract the tops. Keep the bottom. Simplify if you can!

Add: $\frac{1}{6} + \frac{4}{6}$

1. *Same denominator* (6)*. Add numerators:* $1 + 4 = 5$.
2. *Keep the denominator.* $\frac{5}{6}$.
3. *Check: can* $\frac{5}{6}$ *be simplified? No (5 and 6 share no common factor).*

Answer: $\frac{1}{6} + \frac{4}{6} = \frac{5}{6}$

Subtract and Simplify: $\frac{7}{8} - \frac{3}{8}$

1. *Subtract numerators:* $7 - 3 = 4$.
2. *Keep the denominator.* $\frac{4}{8}$.
3. *Simplify:* $\frac{4}{8} = \frac{4 \div 4}{8 \div 4} = \frac{1}{2}$.

Answer: $\frac{7}{8} - \frac{3}{8} = \frac{1}{2}$

Get Online
Find more at
ViewMath.com/TX-Grade4
ViewMath.com

Add and Subtract Fractions — Practice

1. $\frac{2}{5} + \frac{1}{5} =$ ________

2. $\frac{3}{10} + \frac{5}{10} =$ ________

3. $\frac{5}{6} - \frac{2}{6} =$ ________

4. $\frac{9}{12} - \frac{5}{12} =$ ________

5. $\frac{2}{8} + \frac{3}{8} + \frac{1}{8} =$ __________

6. Kim walked $\frac{3}{10}$ of a mile to school and $\frac{4}{10}$ of a mile to the library. How far did she walk in total?

Answer: ______________ miles

Get Online
Find more at
ViewMath.com/TX-Grade4
ViewMath.com

★ 4.5 Add and Subtract Mixed Numbers ★

In this lesson you will learn:

- *Add mixed numbers with like denominators, including regrouping*
- *Subtract mixed numbers with like denominators, including borrowing*
- *Convert between mixed numbers and improper fractions*

Working with Mixed Numbers

*A **mixed number** has a whole part and a fraction part: $3\frac{2}{5}$ means $3 + \frac{2}{5}$.*

Adding Mixed Numbers:

1. *Add the whole numbers.*
2. *Add the fractions.*
3. *If the fraction is ≥ 1, **regroup** (carry 1 to the whole number).*

***Example:** $2\frac{3}{4} + 1\frac{3}{4}$*

Wholes: $2 + 1 = 3$. *Fractions: $\frac{3}{4} + \frac{3}{4} = \frac{6}{4} = 1\frac{2}{4}$.*

Regroup: $3 + 1\frac{2}{4} = 4\frac{2}{4} = 4\frac{1}{2}$.

***Subtracting Mixed Numbers:** If the fraction you're subtracting is bigger, **borrow** 1 from the whole number and add it to the fraction.*

***Example:** $5\frac{1}{3} - 2\frac{2}{3}$*

Can't subtract $\frac{2}{3}$ from $\frac{1}{3}$, so borrow: $5\frac{1}{3} = 4\frac{4}{3}$.

Now: $4\frac{4}{3} - 2\frac{2}{3} = 2\frac{2}{3}$.

TIP

When in doubt, convert both mixed numbers to improper fractions, do the math, then convert back!

Get Online

Find more at
ViewMath.com/TX-Grade4

ViewMath.com

Add: $3\frac{5}{6} + 2\frac{4}{6}$

1. *Add wholes:* $3 + 2 = 5$.
2. *Add fractions:* $\frac{5}{6} + \frac{4}{6} = \frac{9}{6} = 1\frac{3}{6} = 1\frac{1}{2}$.
3. *Regroup:* $5 + 1\frac{1}{2} = 6\frac{1}{2}$.

Answer: $3\frac{5}{6} + 2\frac{4}{6} = 6\frac{1}{2}$

Get Online
Find more at
ViewMath.com/TX-Grade4
ViewMath.com

Mixed Numbers — Practice

1. $1\frac{2}{5} + 3\frac{1}{5} =$ __________

2. $2\frac{3}{8} + 4\frac{7}{8} =$ __________

3. $5\frac{3}{4} - 2\frac{1}{4} =$ __________

4. $6\frac{1}{3} - 3\frac{2}{3} =$ __________

5. Convert $\frac{17}{5}$ to a mixed number. __________

6. A board is $8\frac{1}{4}$ feet long. A carpenter cuts off $3\frac{3}{4}$ feet. How much board is left?

Answer: ______________ feet

★ 4.6 Multiply a Fraction by a Whole Number ★

In this lesson you will learn:

- *Understand fraction multiplication as repeated addition*
- *Multiply a whole number by a fraction using the rule* $n \times \frac{a}{b} = \frac{n \times a}{b}$
- *Convert improper fraction results to mixed numbers*

Multiplying Whole Numbers and Fractions

$n \times \frac{a}{b}$ *means* n ***groups*** *of* $\frac{a}{b}$ *— the same as adding* $\frac{a}{b}$ *repeatedly!*

$$4 \times \frac{2}{5} = \frac{2}{5} + \frac{2}{5} + \frac{2}{5} + \frac{2}{5} = \frac{8}{5}$$

Shortcut Rule: *Multiply the whole number by the* ***numerator****. Keep the* ***denominator*** *the same.*

$$n \times \frac{a}{b} = \frac{n \times a}{b}$$

Example: $4 \times \frac{2}{5} = \frac{4 \times 2}{5} = \frac{8}{5} = 1\frac{3}{5}$

Multiply the top. Keep the bottom. Convert to a mixed number if needed!

“Only the NUMERATOR gets multiplied — the denominator is the name of the pieces and doesn't change!”

Multiply: $6 \times \frac{3}{4}$

1. *Multiply the numerator:* $6 \times 3 = 18$.
2. *Keep the denominator:* $\frac{18}{4}$.
3. *Convert:* $18 \div 4 = 4$ *remainder* 2. *So* $\frac{18}{4} = 4\frac{2}{4} = 4\frac{1}{2}$.

Answer: $6 \times \frac{3}{4} = 4\frac{1}{2}$

Word Problem: Baking Cookies

Each batch of cookies needs $\frac{3}{4}$ cup of sugar. How much sugar do you need for 3 batches?

1. *Write the multiplication:* $3 \times \frac{3}{4}$.
2. *Multiply:* $\frac{3 \times 3}{4} = \frac{9}{4}$.
3. *Convert:* $\frac{9}{4} = 2\frac{1}{4}$.

Answer: *You need* $2\frac{1}{4}$ *cups of sugar.*

Multiply Fractions — Practice

1. $3 \times \frac{2}{5} =$ ________

2. $5 \times \frac{1}{3} =$ ________

3. $4 \times \frac{5}{8} =$ ________

4. $7 \times \frac{2}{6} =$ ________

5. *Write $5 \times \frac{1}{4}$ as a repeated addition and then as a fraction.* ____________________

6. *A caterpillar crawls $\frac{2}{3}$ of a foot each minute. How far does it crawl in 6 minutes?*

Answer: ____________ *feet*

4.7 Tenths and Hundredths

In this lesson you will learn:

- *Convert tenths to equivalent hundredths*
- *Add fractions with denominators of* 10 *and* 100
- *Connect tenths and hundredths to* ***money*** *(dimes and pennies)*

Tenths, Hundredths, and Money

Key relationship: $\frac{1}{10} = \frac{10}{100}$

Each ***tenth*** *equals* 10 ***hundredths****. To convert:*

$$\frac{3}{10} = \frac{3 \times 10}{10 \times 10} = \frac{30}{100}$$

Adding tenths and hundredths: *Convert the tenth to hundredths first, then add.*

$$\frac{4}{10} + \frac{7}{100} = \frac{40}{100} + \frac{7}{100} = \frac{47}{100}$$

Money connection:

- 1 ***dime*** $= \frac{1}{10}$ *of a dollar* $= \frac{10}{100}$ *of a dollar* $= \$0.10$
- 1 ***penny*** $= \frac{1}{100}$ *of a dollar* $= \$0.01$
- $\frac{75}{100}$ *of a dollar* $= 75$ *cents* $= \$0.75$

Money uses hundredths! $\$0.47$ ***means*** $\frac{47}{100}$ ***of a dollar.***

> Think of money! One dime ($\frac{1}{10}$ of a dollar) equals 10 pennies ($\frac{10}{100}$ of a dollar).

Add: $\frac{6}{10} + \frac{23}{100}$

1. Convert: $\frac{6}{10} = \frac{60}{100}$
2. Add: $\frac{60}{100} + \frac{23}{100} = \frac{83}{100}$

Answer: $\frac{6}{10} + \frac{23}{100} = \frac{83}{100}$

Money as Fractions

Write $0.35 as a fraction of a dollar.

1. $\$0.35 = 35$ cents $= 35$ hundredths of a dollar
2. As a fraction: $\frac{35}{100}$
3. Think: 3 dimes ($\frac{3}{10}$) and 5 pennies ($\frac{5}{100}$)

Answer: $\$0.35 = \frac{35}{100}$ of a dollar

Tenths, Hundredths, and Money — Practice

1. $\frac{2}{10} = \frac{?}{100}$? = ________

2. $\frac{5}{10} + \frac{8}{100} =$ ________

3. $\frac{3}{10} + \frac{45}{100} =$ ________

4. Write $0.60 as a fraction of a dollar. ________

5. How many pennies equal $\frac{7}{10}$ of a dollar? ________

6. Jen has $\frac{4}{10}$ of a dollar and finds $\frac{25}{100}$ of a dollar. How much money does she have as a fraction of a dollar?

Answer: ____________ fraction

Get Online

Find more at
ViewMath.com/TX-Grade4

ViewMath.com

★ 4.8 Fractions as Decimals ★

In this lesson you will learn:

- *Convert fractions with denominators of 10 and 100 to decimals*
- *Convert decimals to fractions*
- *Connect fractions and decimals to **money***

Fractions, Decimals, and Money

*A **decimal** uses a dot to show parts of a whole.*

Converting fractions to decimals:

$$\frac{3}{10} = 0.3 \qquad \frac{62}{100} = 0.62 \qquad \frac{4}{100} = 0.04$$

Converting decimals to fractions:

$$0.7 = \frac{7}{10} \qquad 0.35 = \frac{35}{100} \qquad 0.09 = \frac{9}{100}$$

Money connection: *Every dollar amount IS a decimal!*

Money	*Decimal*	*Fraction*
\$0.25	0.25	$\frac{25}{100}$
\$0.50	0.50	$\frac{50}{100}$ *(or* $\frac{5}{10}$*)*
\$0.75	0.75	$\frac{75}{100}$
\$0.10	0.10	$\frac{10}{100}$ *(or* $\frac{1}{10}$*)*

1 digit after the decimal = tenths. 2 digits = hundredths. Money always uses 2 decimal places!

TIP

$\frac{4}{100}$ *is four* ***hundredths****, so it goes in the hundredths place:* 0.04*. The zero in the tenths place is a placeholder!*

Connecting Fractions, Decimals, and Money

Write $\frac{75}{100}$ ***as a decimal and as a dollar amount.***

1. $\frac{75}{100}$*: The 75 fills the tenths and hundredths places* $\rightarrow 0.75$
2. *As money:* $0.75 = \$0.75$ *(seventy-five cents)*

Answer: $\frac{75}{100} = 0.75 = \$0.75$

Fractions as Decimals — Practice

1. $\frac{6}{10} =$ ________

2. $\frac{42}{100} =$ ________

3. $\frac{7}{100} =$ ________

4. $0.9 = \frac{?}{10}$? = ________

5. Write \$0.45 as a fraction of a dollar. __________

6. A toy costs $\frac{80}{100}$ of a dollar. Write this as a decimal and as a dollar amount.

Answer: _______________

4.9 Compare Decimals

In this lesson you will learn:

- *Compare decimals to the hundredths place*
- *Use $>$, $<$, and $=$ to compare decimals*
- *Compare and order **money amounts** using decimal reasoning*

Comparing Decimals and Money

*To compare decimals, line up the decimal points and compare **left to right**:*

Step 1: *Compare the **ones** place.*

Step 2: *If equal, compare the **tenths** place.*

Step 3: *If equal, compare the **hundredths** place.*

Tip: *Add trailing zeros to make both numbers the same length.*

$$0.5 = 0.50 \quad \text{so} \quad 0.5 \text{ and } 0.50 \text{ are EQUAL}$$

Money connection: *Comparing prices works the same way!*

$\$3.45$ *vs.* $\$3.50$*: Ones same (3). Tenths:* $4 < 5$*. So* $\$3.45 < \$3.50$*.*

Don't be tricked! $0.5 > 0.38$ ***because*** 5 ***tenths*** > 3 ***tenths, even though*** $38 > 5$.

"Think about money! Would you rather have $\$0.50$ or $\$0.38$? Of course $\$0.50$ — that's 50 cents vs. 38 cents!"

Compare Prices

A pencil costs $\$0.85$ and an eraser costs $\$0.90$. Which costs more?

1. *Both have $\$0$ in the ones. Compare tenths: 8 vs. 9.*
2. $8 < 9$, *so* $\$0.85 < \$0.90$.

Answer: The eraser ($\$0.90$) costs more.

Compare Decimals — Practice

1. 0.6 _____ 0.60 *(>, <, or =)*

2. 0.35 _____ 0.4 *(>, <, or =)*

3. \$2.75 _____ \$2.57 *(>, <, or =)*

4. \$0.09 _____ \$0.10 *(>, <, or =)*

5. *Order from least to greatest:* \$1.50, \$1.05, \$1.55, \$1.35 ____________________

6. *Sam has* \$0.45 *and Ava has* \$0.54. *Who has more money?*

Answer: ____________ *name*

Get Online

Find more at
ViewMath.com/TX-Grade4

ViewMath.com

★ 4.10 Add and Subtract Decimals to Hundredths ★

In this lesson you will learn:

- *Add and subtract decimals to the **hundredths** place*
- *Line up **decimal points** before computing*
- *Solve money problems using decimal operations*

Line Up the Decimal Points!

*When you add or subtract decimals, stack the numbers so the **decimal points** line up. Then compute from right to left, just like whole numbers.*

Right:

$$\begin{array}{r} 3.45 \\ +\,2.30 \\ \hline 5.75 \end{array}$$

*If one number has fewer decimal places, add **zeros** to fill in:*

$3.4 = 3.40 \qquad 7 = 7.00 \qquad 12.5 = 12.50$

Line up the decimals, add zeros if needed, then compute right to left. Bring the decimal point straight down!

"The #1 rule: LINE UP THE DECIMAL POINTS! Once you do that, it works just like regular addition and subtraction."

Get Online

Find more at
ViewMath.com/TX-Grade4

ViewMath.com

Adding Decimals: $4.63 + 2.85$

1. *Line up decimal points and add right to left.*
2. *Hundredths:* $3 + 5 = 8$.
3. *Tenths:* $6 + 8 = 14$. *Write* 4, *carry* 1.
4. *Ones:* $4 + 2 + 1 = 7$.

Answer: 7.48

Subtracting from a Whole Number: $6 - 2.43$

1. *Rewrite* 6 *as* 6.00.
2. *Subtract:*

$$\begin{array}{r} 6.00 \\ -\ 2.43 \\ \hline 3.57 \end{array}$$

Answer: 3.57

Add and Subtract Decimals — Practice

1. $3.24 + 5.63 =$ ________

2. $7.85 + 4.39 =$ ________

3. $9.54 - 3.21 =$ ________

4. $10.00 - 4.35 =$ ________

5. $6.5 + 3.78 =$ ________

6. *Ava buys a notebook for $\$3.95$ and a pen for $\$1.28$. How much does she spend in total?*

Answer: ____________ dollars

CHAPTER

5

Measurement & Data

What's Inside

★ 5.1 Metric Units and Conversions ★

In this lesson you will learn:

- *Know the metric units for length, mass, and liquid volume*
- *Convert larger metric units to smaller ones by multiplying*
- *Use key facts: $1\ km = 1{,}000\ m$, $1\ m = 100\ cm$, $1\ kg = 1{,}000\ g$, $1\ L = 1{,}000\ mL$*

Metric Conversions — Multiply by 10, 100, or 1,000

The metric system is built on powers of 10. To go from a **bigger** unit to a **smaller** unit, **multiply**.

Length:

- $1\ km = 1{,}000\ m$
- $1\ m = 100\ cm$
- $1\ cm = 10\ mm$

Mass: $1\ kg = 1{,}000\ g$

Liquid Volume: $1\ L = 1{,}000\ mL$

Bigger unit → smaller unit = MULTIPLY.

Smaller unit → bigger unit = DIVIDE.

> “Think "King Henry Died by Drinking Chocolate Milk" — Kilo, Hecto, Deka, Base, Deci, Centi, Milli. Each step is $\times 10$!”

Convert 3 km to meters

1. $1\ km = 1{,}000\ m$
2. $3\ km = 3 \times 1{,}000 = 3{,}000\ m$

Answer: $3\ km =$ $3{,}000\ m$

Convert 5 L to milliliters

1. $1\ L = 1{,}000\ mL$
2. $5\ L = 5 \times 1{,}000 = 5{,}000\ mL$

Answer: $5\ L =$ $5{,}000\ mL$

Metric Units and Conversions — Practice

1. $7\ m =$ ________ cm
2. $4\ km =$ ________ m
3. $9\ kg =$ ________ g
4. $2\ L =$ ________ mL
5. $6\ m =$ ________ mm (Hint: $1\ m = 1{,}000\ mm$)
6. A water bottle holds $500\ mL$. How many milliliters are in 3 bottles?

Answer: ____________ mL

★ 5.2 Customary Units and Conversions ★

In this lesson you will learn:

- *Know the customary units for length, weight, and liquid volume*
- *Convert larger customary units to smaller ones by multiplying*
- *Memorize key facts: 1 ft $= 12$ in, 1 yd $= 3$ ft, 1 lb $= 16$ oz, 1 gal $= 4$ qt*

Customary Unit Conversions

In the United States, we use the **customary system**. *The conversion factors are not as neat as the metric system, so memorize these key facts:*

Length:

- 1 foot (ft) $= 12$ inches (in)
- 1 yard (yd) $= 3$ feet (ft) $= 36$ inches
- 1 mile (mi) $= 5{,}280$ feet

Weight:

- 1 pound (lb) $= 16$ ounces (oz)

Liquid Volume:

- 1 gallon (gal) $= 4$ quarts (qt)
- 1 quart $= 2$ pints (pt)
- 1 pint $= 2$ cups (c)

Bigger unit → smaller unit = MULTIPLY. Same rule as metric!

“Gallon man” has 4 Quarts, each Quart has 2 Pints, each Pint has 2 Cups. So 1 gallon $= 16$ cups!

Convert 5 feet to inches

1. 1 *ft* $= 12$ *in*
2. 5 *ft* $= 5 \times 12 = 60$ *in*

Answer: 5 *ft* $=$ 60 *in*

Convert 3 gallons to quarts

1. 1 *gal* $= 4$ *qt*
2. 3 *gal* $= 3 \times 4 = 12$ *qt*

Answer: 3 *gal* $=$ 12 *qt*

Customary Units and Conversions – Practice

1. $4\ ft =$ ________ in

2. $2\ yd =$ ________ ft

3. $3\ lb =$ ________ oz

4. $5\ gal =$ ________ qt

5. $7\ qt =$ ________ pt

6. A recipe needs 2 cups of milk. Jake wants to make 4 batches. How many cups does he need? How many pints is that?

Answer: ____________ cups; pints

5.3 Measurement Word Problems

In this lesson you will learn:

- *Solve word problems involving measurement conversions*
- *Solve **elapsed time** problems (hours, minutes, seconds)*
- *Solve **money** word problems (making change, total cost)*

Measurement, Time, and Money Word Problems

Measurement problems: *Convert units, then add, subtract, or compare.*

Time conversions:

- 1 *hour* $= 60$ *minutes* $\quad$ 1 *minute* $= 60$ *seconds*
- ***Elapsed time*** $=$ *End time* $-$ *Start time*

Example: *Start at* $2{:}15$ *PM, end at* $4{:}45$ *PM.*

$4{:}45 - 2{:}15 = 2$ *hours* 30 *minutes.*

Money problems:

- ***Making change:*** *Amount paid* $-$ *Cost* $=$ *Change*
- ***Total cost:*** *Price* $\times$ *Quantity* $=$ *Total*

Read the problem carefully. Decide: Do I add, subtract, multiply, or convert?

"For elapsed time, count forward on a clock. Jump to the next hour first, then count the extra minutes!"

Elapsed Time

A movie starts at* 3:20 *PM and ends at* 5:05 *PM. How long is the movie?

1. *From* 3:20 *to* 5:20 = 2 *hours*
2. *From* 5:20 *back to* 5:05 = −15 *minutes*
3. *Total:* 2 *hours* −15 *minutes* = 1 *hour* 45 *minutes*

Answer: 1 *hour* 45 *minutes*

Money: Making Change

You buy a notebook for* \$3.75 *and pay with a* \$5 *bill. How much change?

1. *Change* = \$5.00 − \$3.75
2. \$5.00 − \$3.75 = \$1.25

Answer: \$1.25 *change*

Measurement Word Problems — Practice

1. *How many minutes are in 3 hours?* __________
2. *A class starts at 9:15 AM and ends at 10:00 AM. How long is the class?* __________
3. *You buy a pen for $1.50 and a folder for $2.25. What is the total?* __________
4. *You pay $10.00 for items costing $7.35. What is your change?* __________
5. *A runner starts at 4:30 PM and runs for 1 hour 20 minutes. What time does she finish?* __________
6. *Sam has 5 meters of ribbon. He uses 250 cm on a project. How many centimeters of ribbon are left?*

Answer: ______________

★ 5.4 Area of Rectangles ★

◎ In this lesson you will learn:

- *Use the formula $A = l \times w$ to find the area of a rectangle*
- *Find the area of composite shapes made of rectangles*
- *Solve real-world area problems with correct square units*

The Area Formula

***Area** is the amount of space inside a flat shape, measured in **square units** (sq in, sq cm, sq ft, etc.).*

For any rectangle:

$$A = l \times w$$

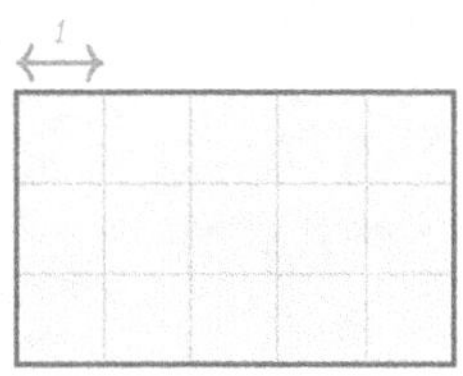

3 *rows* $\times$ 5 *columns* $= 15$ *square units.*

Area = length × width. Always use SQUARE units in your answer!

Get Online
Find more at
ViewMath.com/TX-Grade4
ViewMath.com

Find the area of a rectangle with $l = 8$ cm and $w = 6$ cm

1. Use the formula: $A = l \times w$
2. $A = 8 \times 6 = 48$ sq cm

Answer: $A = 48$ **sq cm**

Composite Shape

An L-shaped room is made of two rectangles. One is 4 m $\times$ 3 m. The other is 2 m $\times$ 5 m. What is the total area?

1. Area of rectangle 1: $4 \times 3 = 12$ sq m
2. Area of rectangle 2: $2 \times 5 = 10$ sq m
3. Total area: $12 + 10 = 22$ sq m

Answer: **Total area** $= 22$ **sq m**

Get Online
Find more at
ViewMath.com/TX-Grade4
ViewMath.com

Area of Rectangles — Practice

1. $l = 7$ *in*, $w = 4$ *in.*

$A =$ __________

2. $l = 12$ *cm*, $w = 5$ *cm.*

$A =$ __________

3. $l = 9$ *ft*, $w = 9$ *ft.*

$A =$ __________

4. $l = 15$ *m*, $w = 3$ *m.*

$A =$ __________

5. *A rectangle has an area of* 36 *sq cm and a length of* 9 *cm. What is the width?* __________

6. *A garden is* 10 *feet long and* 6 *feet wide. Each bag of mulch covers* 12 *square feet. How many bags are needed?*

Answer: ______________ *bags*

★ 5.5 Perimeter of Rectangles ★

In this lesson you will learn:

- Use the formula $P = 2l + 2w$ to find the perimeter of a rectangle
- Find a missing side length when the perimeter is known
- Solve real-world perimeter problems

The Perimeter Formula

Perimeter is the total distance **around** a shape.

A rectangle has 2 long sides and 2 short sides:

$$P = 2l + 2w$$

Example: A rectangle with $l = 8$ cm and $w = 3$ cm:

$$P = 2(8) + 2(3) = 16 + 6 = 22 \text{ cm}$$

Perimeter is measured in regular units (cm, in, ft) — NOT square units!

“Think of perimeter as walking around the outside of a shape. Area fills the inside; perimeter goes around the outside!”

Find the perimeter: $l = 12$ *in*, $w = 5$ *in*

1. $P = 2l + 2w$
2. $P = 2(12) + 2(5) = 24 + 10 = 34$ *in*

Answer: $P = 34$ *in*

Find the Missing Side

A rectangle has a perimeter of 30 cm and a length of 10 cm. What is the width?

1. $P = 2l + 2w$, so $30 = 2(10) + 2w$
2. $30 = 20 + 2w$
3. $2w = 10$, so $w = 5$ *cm*

Answer: $w = 5$ *cm*

Get Online
Find more at
ViewMath.com/TX-Grade4
ViewMath.com

Perimeter of Rectangles — Practice

1. $l = 9$ ft, $w = 4$ ft.

   $P =$ __________

2. $l = 11$ cm, $w = 7$ cm.

   $P =$ __________

3. $l = 6$ m, $w = 6$ m.

   $P =$ __________

4. $P = 40$ in, $l = 14$ in.

   $w =$ __________

5. True or False: A square with side 5 cm has a perimeter of 25 cm. __________

6. Mia wants to put a fence around a rectangular garden that is 15 feet long and 8 feet wide. How many feet of fencing does she need?

   Answer: ______________ ft

5.6 Line Plots with Fractions

In this lesson you will learn:

- *Read and interpret data from a line plot with fractional measurements*
- *Create a line plot using data in halves, quarters, or eighths*
- *Solve addition and subtraction problems using line plot data*

What Is a Line Plot?

*A **line plot** uses a number line with X marks above it. Each X represents one data point.*

Example: *Students measured crayon lengths (in inches):*

$\frac{1}{4}, \frac{1}{2}, \frac{1}{4}, \frac{3}{4}, \frac{1}{2}, 1, \frac{1}{2}, \frac{3}{4}, \frac{1}{4}, \frac{1}{2}$

X	X		
X	X	X	
X	X	X	X
$\frac{1}{4}$	$\frac{1}{2}$	$\frac{3}{4}$	1

From this plot: 3 *crayons are* $\frac{1}{4}$ *in,* 4 *crayons are* $\frac{1}{2}$ *in,* 2 *are* $\frac{3}{4}$ *in, and* 1 *is* 1 *in.*

Count the X's above each value to find how many data points are at that measurement.

Using a Line Plot

Using the crayon data above, what is the total length of the 3 shortest crayons?

1. *The 3 shortest crayons are all $\frac{1}{4}$ in.*
2. *Total: $\frac{1}{4}+\frac{1}{4}+\frac{1}{4}=\frac{3}{4}$ in*

Answer: $\frac{3}{4}$ inch

Line Plots with Fractions — Practice

Use this line plot of plant heights (in inches) for Problems 1–5.

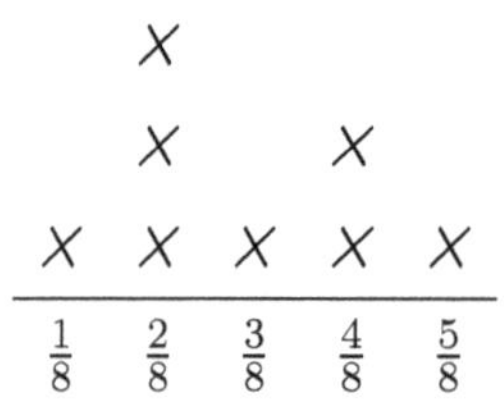

1. How many plants are $\frac{2}{8}$ inches tall? __________
2. How many plants were measured in all? __________
3. What is the most common height? __________
4. What is the difference between the tallest and shortest plant? __________
5. What is the total height of the two tallest plants? __________

5.7 What Is an Angle?

In this lesson you will learn:

- Understand that an angle is formed by two rays sharing a common endpoint
- Know that a full turn is $360°$ and one degree is $\frac{1}{360}$ of a full turn
- Identify angles in everyday objects

Angles — Turns and Degrees

An **angle** is formed when two **rays** meet at a common point called the **vertex**.

Think of a clock: the two hands are rays, and the center pin is the vertex.

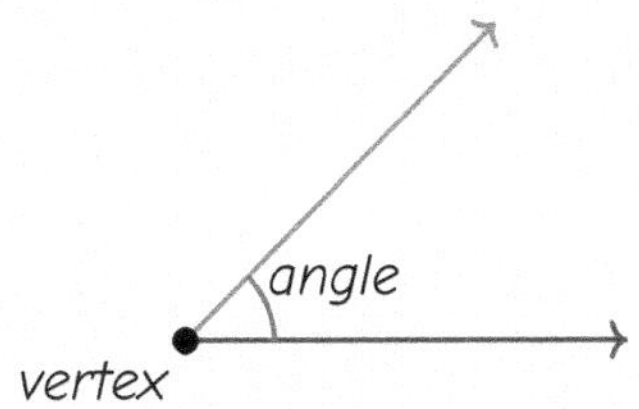

Key angle amounts:

- **Full turn** $= 360°$ (all the way around)
- **Half turn** $= 180°$ (a straight line)
- **Quarter turn** $= 90°$ (a right angle, like a corner)

One degree ($1°$) is $\frac{1}{360}$ of a full turn around a circle.

"Open a book partway — you just made an angle! The wider you open it, the bigger the angle."

How Many Degrees?

A door opens from closed ($0°$) to fully open against the wall. How many degrees did it turn?

1. *A closed door and the wall make a straight line when fully open.*
2. *A straight line = half a turn = $180°$.*

Answer: The door turned $180°$.

What Is an Angle? — Practice

1. *How many degrees are in a full turn?* ________

2. *How many degrees are in a right angle?* ________

3. *How many right angles make a full turn?* ________

4. *True or False: A straight line forms a 180° angle.* ________

5. *A pizza is cut into 4 equal slices. What angle does each slice make at the center?* ________

6. *A Ferris wheel makes $\frac{1}{4}$ of a full turn. How many degrees has it turned?*

Answer: ____________ *degrees*

5.8 Measure and Draw Angles

In this lesson you will learn:

- *Use a protractor to measure angles in whole-number degrees*
- *Draw an angle of a given degree measure*
- *Read the correct scale on a two-scale protractor*

Using a Protractor

*A **protractor** is a half-circle tool used to measure and draw angles. It has two scales (inner and outer) that go in opposite directions.*

How to measure an angle:

1. *Place the protractor's **center mark** on the **vertex**.*
2. *Line up one ray with the $0°$ line (the baseline).*
3. *Read where the **other ray** crosses the scale.*
4. *Use the scale that starts at $0°$ where your first ray is!*

How to draw an angle:

1. *Draw one ray. Place the protractor on the endpoint.*
2. *Find the desired degree on the scale and make a dot.*
3. *Remove the protractor and connect the endpoint to the dot.*

***Check your work:** If the angle looks small (acute), your answer should be less than $90°$. If it looks wide (obtuse), it should be more than $90°$.*

"A protractor has TWO scales. Always start from $0°$ and count up. If the number seems wrong, try the other scale!"

Get Online
Find more at
ViewMath.com/TX-Grade4
ViewMath.com

Estimate Then Measure

Estimate: Does a corner of a book look like $45°$, $90°$, or $135°$?

1. *A book corner forms a perfect square corner.*
2. *A square corner is a **right angle** $= 90°$.*
3. *Measuring with a protractor confirms: $90°$.*

Answer: $90°$

Measure and Draw Angles — Practice

1. *An angle measures 60°. Is it acute, right, or obtuse?* __________

2. *An angle measures 135°. Is it acute, right, or obtuse?* __________

3. *True or False: A protractor can measure angles up to 360°.* __________

4. *The hands of a clock at 3:00 form what angle?* __________

5. *The hands of a clock at 6:00 form what angle?* __________

6. *Jada measured an angle and got 150°. Her friend got 30° for the same angle. Who is correct if the angle looks wide (obtuse)?*

Answer: ______________ *name*

★ 5.9 Add and Subtract Angles ★

In this lesson you will learn:

- *Understand that angle measurement is additive*
- *Find the total when two or more angles share a side*
- *Subtract to find an unknown angle measure*

Angle Addition

When a big angle is split into two smaller angles (no gaps, no overlaps):

$$\textit{Big angle} = \textit{Angle 1} + \textit{Angle 2}$$

Example: *A ray splits a* 90° *angle into two parts. One part is* 55°.

$$\textit{Other part} = 90 - 55 = 35^\circ$$

Check: $55 + 35 = 90$ ✓

This works with straight angles too:

$$\textit{Straight angle} = 180^\circ$$

If one part is 120°, *the other is* $180 - 120 = 60^\circ$.

Angles add up like numbers. If you know the whole and one part, subtract to find the missing part!

"Think of cutting a pie: the two slices always add up to the whole pie angle. If you know one slice, subtract to find the other!"

Find the Missing Angle

A right angle ($90°$) is split into two angles. One measures $40°$. Find the other.

1. $90 = 40 + ?$
2. $? = 90 - 40 = 50°$

Answer: $50°$

Add Three Angles

Three angles along a straight line measure $45°$, $75°$, and ?. Find the missing angle.

1. A straight line $= 180°$
2. $45 + 75 = 120$
3. $? = 180 - 120 = 60°$

Answer: $60°$

Add and Subtract Angles — Practice

1. *Two angles share a side. One is 65° and the other is 25°. What is the total?* __________

2. *A 180° angle is split into two parts. One part is 110°. What is the other?* __________

3. *A 90° angle is split into two parts. One part is 38°. What is the other?* __________

4. $45^\circ + 45^\circ + ? = 180^\circ$. *Find* ?. __________

5. *True or False: If two angles add up to 90°, they form a right angle.* __________

6. *A pie is cut into two pieces. The bigger piece has an angle of 230° at the center. What is the angle of the smaller piece?*

Answer: ______________ *degrees*

5.10 Personal Financial Literacy

In this lesson you will learn:

- Tell the difference between **fixed** and **variable** expenses
- Calculate simple **profit** (income minus expenses)
- Create a simple **budget** to plan spending and saving

Profit, Expenses, and Budgets

Fixed expenses *stay the same each month (rent, phone bill).*

Variable expenses *change each month (groceries, fun activities).*

Profit *is what you keep after paying costs:*

$$\textit{Profit} = \textit{Income} - \textit{Expenses}$$

If income > expenses → profit! *If expenses > income → loss.*

A ***budget*** *is a plan:*

$$\textit{Income} = \textit{Spending} + \textit{Saving}$$

If you get \$20 allowance, you might plan: \$12 spending + \$8 saving.

Profit = Income − Expenses. Always subtract ALL your expenses!

"Understanding money isn't just about counting coins — it's about making smart choices with saving, spending, and planning!"

Get Online
Find more at
ViewMath.com/TX-Grade4

ViewMath.com

Calculating Profit

Jake sells 8 bracelets for $\$3.00$ each. He spent $\$12.00$ on beads and string. What is his profit?

1. *Income:* $8 \times \$3.00 = \$24.00$.
2. *Expenses:* $\$12.00$.
3. *Profit:* $\$24.00 - \$12.00 = \$12.00$.

Answer: $\$12.00$ profit

Building a Budget

Emma earns $\$30$ per month. Expenses: $\$5$ snacks, $\$8$ art supplies, $\$4$ toys. She saves $\$10$. Can she afford it?

1. *Total expenses:* $\$5 + \$8 + \$4 = \$17$.
2. *Expenses + savings* $= \$17 + \$10 = \$27$.
3. *Income is $\$30$. Since $\$30 > \$27$, yes — with $\$3$ left over!*

Answer: Yes! She can afford everything and save $\$10$.

Financial Literacy — Practice

1. *Monthly piano lessons: $40 every month. Fixed or variable?* __________

2. *Grocery bill: $250 one month, $310 the next. Fixed or variable?* __________

3. *Maria buys 10 candy bars for $0.50 each and sells them for $1.00 each. What is her profit?*

   Answer: ______________ dollars

4. *Ben spends $15 on supplies for a bake sale. He earns $42. What is his profit?*

   Answer: ______________ dollars

5. *Leo gets $25 allowance. He saves $8, spends $10 on games and $5 on snacks. How much is left?*

   Answer: ______________ dollars

6. *True or false: Profit means you earned more money than you spent.* __________

★ 5.11 Dot Plots and Stem-and-Leaf Plots ★

In this lesson you will learn:

- *Create and read **dot plots** (line plots)*
- *Create and read **stem-and-leaf plots***
- *Interpret data and find mode, range, and median from each display*

Dot Plots and Stem-and-Leaf Plots

*A **dot plot** places one dot for each data value above a number line. Stacked dots show how often a value appears.*

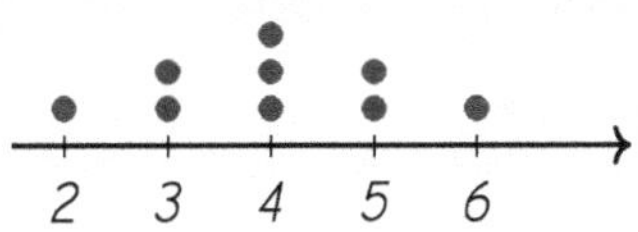

*A **stem-and-leaf plot** splits each number into a **stem** (leading digits) and **leaf** (last digit). Leaves must be in order!*

Stem	Leaves
7	2 5 8
8	1 3 5 5
9	0 2 5

Key: $7|2$ *means* 72.

Count the total number of leaves to find how many data points there are!

Get Online
Find more at
ViewMath.com/TX-Grade4
ViewMath.com

> *"A dot plot uses dots; a stem-and-leaf plot uses actual digits. Both help you see the "shape" of the data at a glance!"*

Reading a Stem-and-Leaf Plot

Stem	Leaves
2	1 4 7
3	0 3 3 6 9
4	2 5

Key: $3|6 = 36$. *Find the total count, mode, and range.*

1. *Count all leaves:* $3 + 5 + 2 = 10$ *data values.*
2. *Mode:* 33 *(leaf* 3 *appears twice in stem* 3*).*
3. *Range:* $45 - 21 = 24$.

Answer: 10 **values. Mode** $=$ 33. **Range** $=$ 24.

Dot Plots and Stem-and-Leaf — Practice

Use this stem-and-leaf plot (minutes to complete a puzzle):

Stem	Leaves
1	0 2 5 8
2	1 3 3 6 7
3	0 4

Key: $2|3 = 23$ *minutes*

1. *How many people completed the puzzle?* __________
2. *What was the fastest time?* __________
3. *What is the mode?* __________
4. *What is the range?* __________
5. *True or false: In a stem-and-leaf plot, leaves go from greatest to least.* __________
6. *True or false: Each dot on a dot plot represents one data point.* __________

★ 5.12 Time Extensions: Elapsed Time and Conversions ★

◎ In this lesson you will learn:

- Convert between seconds, minutes, hours, and days
- Calculate **elapsed time** between two events
- Understand US **time zones** and how they affect the clock

Time Conversions, Elapsed Time, and Time Zones

Key facts:

$$1 \text{ minute} = 60 \text{ seconds}$$

$$1 \text{ hour} = 60 \text{ minutes}$$

$$1 \text{ day} = 24 \text{ hours}$$

Larger → smaller: MULTIPLY. Smaller → larger: DIVIDE.

Elapsed time = time that passes between start and end. Count forward:

1. Count up to the next full hour.
2. Count the remaining hours.
3. Add leftover minutes.

US Time Zones (each 1 hour apart): Eastern → Central → Mountain → Pacific.

Going west → subtract hours. Going east → add hours.

Time uses base 60, not base 100! When minutes reach 60, trade for 1 hour.

> *"If it's 3:00 PM in New York (Eastern), it's 12:00 PM in Los Angeles (Pacific) — 3 hours behind!"*

Finding Elapsed Time

Soccer practice: 3:45 PM to 5:20 PM. How long?

1. $3{:}45$ *to* $4{:}00 = 15$ *minutes.*
2. $4{:}00$ *to* $5{:}00 = 1$ *hour.*
3. $5{:}00$ *to* $5{:}20 = 20$ *minutes.*
4. *Total:* 1 *hour* $+15 + 20 = 1$ *hour* 35 *minutes.*

Answer: 1 hour and 35 minutes

Time Zone Problem

Grandma in New York calls at 7:00 PM Eastern. What time is it in Los Angeles?

1. *Pacific is 3 hours behind Eastern.*
2. $7{:}00$ *PM* -3 *hours* $= 4{:}00$ *PM.*

Answer: 4:00 PM Pacific Time

Time Extensions — Practice

1. 4 hours = __________ minutes

2. 300 seconds = __________ minutes

3. An art class starts at 10:15 AM and ends at 11:45 AM. How long is the class?

Answer: ______________ time

4. Emma starts homework at 4:30 PM and works for 1 hour and 25 minutes. What time does she finish?

Answer: ______________ time

5. It is 5:00 PM in New York (Eastern). What time is it in Chicago (Central)? __________

6. True or false: 2 hours = 100 minutes. __________

CHAPTER

6

Geometry

What's Inside

6.1 Lines, Rays, and Segments

In this lesson you will learn:

- Identify and draw **points**, **lines**, **line segments**, and **rays**
- Understand the differences between a line, a ray, and a segment
- Use correct notation: $\overleftrightarrow{AB}$, $\overrightarrow{AB}$, $\overline{AB}$

Points, Lines, Rays, and Segments

Geometry starts with four basic building blocks:

Point — an exact location, drawn as a dot, named with a capital letter.

Line — goes on **forever in both directions**. Arrows on both ends. Written $\overleftrightarrow{AB}$.

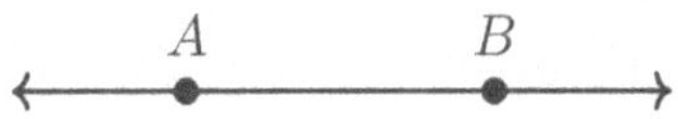

Line Segment — a piece of a line with **two endpoints**. You can measure it! Written $\overline{CD}$.

Ray — starts at **one endpoint** and goes on forever in one direction. Written $\overrightarrow{EF}$.

Line = 0 endpoints. Ray = 1 endpoint. Segment = 2 endpoints.

"Think of a flashlight beam — it starts at a point and goes on forever in one direction. That's a ray!"

Name Each Figure

Look at each figure and decide: line, ray, or segment?

1. *A straight path with arrows on BOTH ends.*
2. *A straight path with a dot on one end and an arrow on the other.*
3. *A straight path with dots on BOTH ends.*

1. *Both arrows → goes forever both ways → **line**.*
2. *One dot + one arrow → starts at a point, goes forever → **ray**.*
3. *Two dots → starts and stops → **line segment**.*

Answer: 1. Line 2. Ray 3. Line Segment

Find Segments in a Triangle

How many line segments does a triangle have?

1. *A triangle has 3 sides.*
2. *Each side starts at one vertex and ends at another vertex — two endpoints.*
3. *Each side is a **line segment**.*

Answer: A triangle has 3 line segments.

Lines, Rays, and Segments — Practice

1. How many endpoints does a ray have? __________
2. How many endpoints does a line segment have? __________
3. True or false: A line has a definite length that you can measure. __________
4. A rectangle has __________ line segments (sides).
5. Write the correct notation for a line through points M and N. __________
6. A pentagon has 5 sides. How many line segments form the pentagon? __________

★ 6.2 Angle Types ★

In this lesson you will learn:

- *Classify angles as **acute**, **right**, **obtuse**, or **straight***
- *Identify angle types in shapes and everyday objects*
- *Estimate angle sizes by comparing to a right angle (90°)*

The Four Angle Types

*An **angle** is formed where two rays share a common endpoint (the **vertex**).*

Type	Measure	How to Remember
Acute	less than 90°	"A-cute" little angle — small!
Right	exactly 90°	A perfect "L" shape.
Obtuse	between 90° and 180°	Big and wide open.
Straight	exactly 180°	Flat like a line.

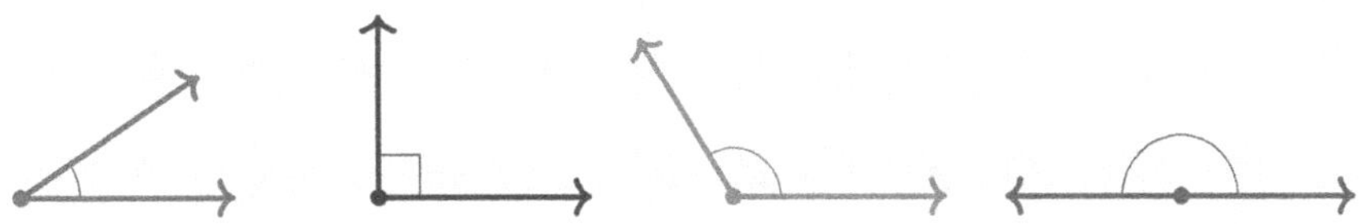

Acute   Right   Obtuse   Straight

Quick test: *Use the corner of a piece of paper (90°) to compare. Smaller than the corner = acute. Matches the corner = right. Bigger than the corner = obtuse.*

> "Acute" sounds like "a cute" — think of a small, cute angle! Obtuse angles are the opposite — big and wide.

Classify Each Angle

What type of angle is each one?

1. *An angle that measures* $55°$.
2. *An angle that measures* $90°$.
3. *An angle that measures* $140°$.

1. $55° < 90° \rightarrow$ ***acute***.
2. $90°$ *exactly* $\rightarrow$ ***right***.
3. $90° < 140° < 180° \rightarrow$ ***obtuse***.

Answer: *1. Acute   2. Right   3. Obtuse*

Angles in a Rectangle

What types of angles does a rectangle have?

1. *A rectangle has* 4 *corners.*
2. *Each corner is a perfect square corner* $= 90°$.
3. *All* 4 *angles are* ***right angles***.

Answer: *All* 4 *angles are right angles* ($90°$).

Angle Types — Practice

1. *Classify* $72°$*:* __________

2. *Classify* $90°$*:* __________

3. *Classify* $115°$*:* __________

4. *Classify* $180°$*:* __________

5. *A triangle has angles of* $60°$, $90°$, *and* $30°$. *Which angle is acute, right, or obtuse?* ____________________

6. *True or false: An obtuse angle is bigger than a right angle.* __________

★ 6.3 Parallel and Perpendicular Lines ★

In this lesson you will learn:

- *Recognize and draw* ***parallel lines*** *and* ***perpendicular lines***
- *Identify parallel and perpendicular lines in shapes and the real world*
- *Use the symbols* $\parallel$ *(parallel) and* $\perp$ *(perpendicular)*

Parallel Lines vs. Perpendicular Lines

Parallel lines *go in the same direction and* ***never cross***. *They stay the same distance apart.*

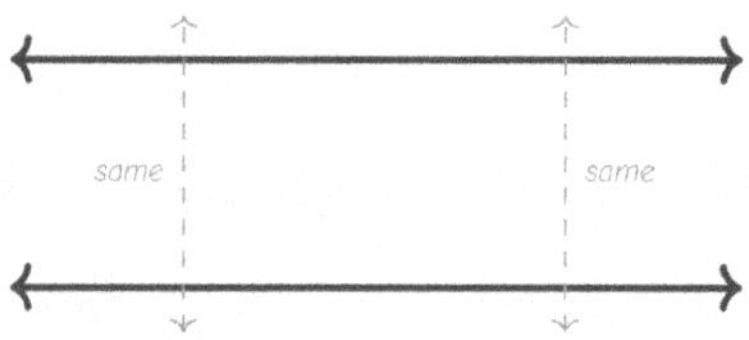

$l \parallel m$

Perpendicular lines *cross each other and form a* ***right angle*** *(90°).*

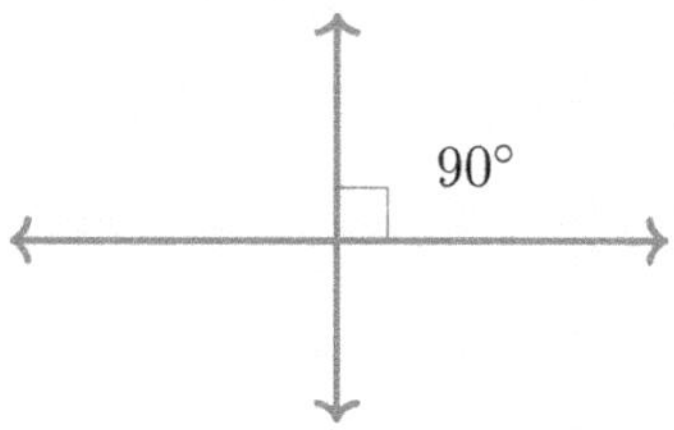

$a \perp b$

Parallel = same direction, never meet. Perpendicular = cross at 90°.

“Railroad tracks are parallel — they never meet! A plus sign (+) shows perpendicular lines — they cross at right angles.”

Parallel or Perpendicular?

Look at the letter "H." Name one pair of parallel line segments and one pair of perpendicular line segments.

1. *The two vertical strokes of H go in the same direction and never cross $\rightarrow$* ***parallel****.*
2. *The horizontal bar meets each vertical stroke at $90°$ $\rightarrow$* ***perpendicular****.*

Answer:** **The two vertical strokes are parallel. The horizontal bar is perpendicular to each vertical stroke.

Lines in a Rectangle

Does a rectangle have parallel and perpendicular sides?

1. *The top and bottom sides go the same direction $\rightarrow$ parallel.*
2. *The left and right sides go the same direction $\rightarrow$ parallel.*
3. *Each corner is $90°$, so every pair of touching sides is perpendicular.*

Answer:** **Yes — 2 pairs of parallel sides and 4 pairs of perpendicular sides.

Parallel and Perpendicular Lines – Practice

1. True or false: Parallel lines eventually cross if you extend them far enough. __________
2. True or false: Perpendicular lines form right angles where they meet. __________
3. How many pairs of parallel sides does a square have? __________
4. How many pairs of perpendicular sides does a square have? __________
5. Name a real-world example of parallel lines. _______________
6. The letter "T" has one pair of perpendicular segments. True or false? __________

Get Online
Find more at
ViewMath.com/TX-Grade4

ViewMath.com

6.4 Classify Shapes

In this lesson you will learn:

- *Classify triangles by their angles (right, acute, obtuse)*
- *Classify quadrilaterals in a **hierarchy** (trapezoid → parallelogram → rectangle/rhombus → square)*
- *Name polygons through **decagon** (10 sides)*

Polygons, Triangles, and the Quadrilateral Hierarchy

*A **polygon** is a closed shape with straight sides.*

Sides	Name
3	Triangle
4	Quadrilateral
5	Pentagon
6	Hexagon
8	Octagon
10	Decagon

Triangle types by angles:

- **Right triangle** — *has one* $90°$ *angle*
- **Acute triangle** — *all angles less than* $90°$
- **Obtuse triangle** — *has one angle greater than* $90°$

Quadrilateral hierarchy *(every shape below is also the shape above it):*

Quadrilateral

↓

Trapezoid (at least 1 pair of parallel sides)

↓

Parallelogram (2 pairs of parallel sides)

↙ ↘

Rectangle (4 right angles) *Rhombus (all sides equal)*

↘ ↙

Square

(4 right angles AND all sides equal)

A square is BOTH a rectangle AND a rhombus. All squares are rectangles, but not all rectangles are squares!

“Think of shape families like a family tree. A square is a special rectangle, and a rectangle is a special parallelogram!”

Name That Shape!

A shape has 4 sides, 2 pairs of parallel sides, and 4 right angles, but NOT all sides are equal.

1. *4 sides → quadrilateral.*
2. *2 pairs of parallel sides → parallelogram.*
3. *4 right angles → rectangle.*
4. *Sides NOT all equal → not a square or rhombus.*

Answer: Rectangle

Triangle Classification

A triangle has angles of $40°$, $110°$, and $30°$. Classify it.

1. *Check: is any angle $= 90°$? No.*
2. *Is any angle $> 90°$? Yes — $110° > 90°$.*

Answer: Obtuse triangle

Classify Shapes – Practice

1. *A shape has 4 equal sides and 4 right angles. What is it?* ___________
2. *A quadrilateral has exactly 1 pair of parallel sides. What is it?* ___________
3. *True or false: A square is always a rectangle.* ___________
4. *A triangle has angles of $60°$, $70°$, and $50°$. What type is it?* ___________
5. *How many sides does a hexagon have?* ___________
6. *True or false: Every rectangle is also a parallelogram.* ___________

★ 6.5 Symmetry ★

In this lesson you will learn:

- *Understand what a **line of symmetry** is*
- *Identify line-symmetric figures and draw lines of symmetry*
- *Determine how many lines of symmetry a figure has*

Lines of Symmetry

*A **line of symmetry** divides a shape into two halves that are **mirror images** of each other. If you fold the shape along the line, both halves match up perfectly.*

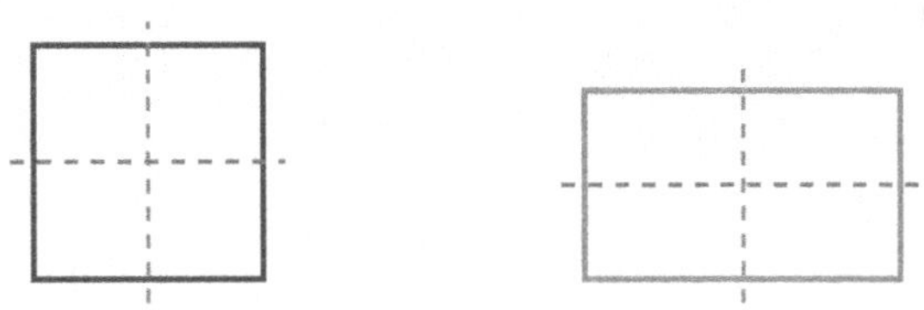

Square: 4 lines　　*Rectangle: 2 lines*

How many lines of symmetry?

Shape	*Lines of Symmetry*
Scalene triangle (no equal sides)	0
Isosceles triangle (2 equal sides)	1
Equilateral triangle (3 equal sides)	3
Rectangle (not square)	2
Square	4
Regular hexagon	6
Circle	*Infinite*

A regular polygon has as many lines of symmetry as it has sides!

“Think of a butterfly — its left wing is a mirror image of its right wing. The body is the line of symmetry!”

Lines of Symmetry in a Square

How many lines of symmetry does a square have?

1. *Fold top to bottom (horizontal line) — both halves match. That's* 1.
2. *Fold left to right (vertical line) — both halves match. That's* 2.
3. *Fold corner to corner (diagonal) — both halves match. That's* 3.
4. *Fold the other diagonal — both halves match. That's* 4.

Answer: A square has 4 lines of symmetry.

Is This Shape Symmetric?

Does the letter "B" have a line of symmetry?

1. *Try a vertical fold — left and right don't match (the bumps are only on the right).*
2. *Try a horizontal fold — the top half mirrors the bottom half!*

Answer: Yes — one horizontal line of symmetry.

Symmetry – Practice

1. *How many lines of symmetry does an equilateral triangle have?* ________
2. *How many lines of symmetry does a rectangle (not a square) have?* ________
3. *True or false: A circle has exactly 4 lines of symmetry.* ________
4. *How many lines of symmetry does a regular pentagon (5 equal sides) have?* ________
5. *Does the letter "A" have a line of symmetry? If yes, how many?* ________
6. *Draw or describe where the line of symmetry is on a heart shape.* ____________

Answer Key

Check Your Answers!

Try each problem first, then look here to check your work.
It's OK to make mistakes — that's how we learn

Chapter 1

Section 1.1

1. 18
2. 28
3. 40
4. 7
5. 24 *cards*
6. 63 *feet*

Explanations

1. *Multiply the comparison factor by the starting amount:* $6 \times 3 = 18$.
2. *Multiply the comparison factor by the starting amount:* $4 \times 7 = 28$.
3. *Multiply the comparison factor by the starting amount:* $8 \times 5 = 40$.
4. *Undo multiplication with division:* $21 \div 3 = 7$, *so the missing factor is* 7.
5. *Luis has* 4 *groups of Ben's* 6 *cards:* $4 \times 6 = 24$ *cards.*

6 *The building is 7 groups of 9 feet:* $7 \times 9 = 63$ *feet.*

> Section 1.2

1 42 stickers | 2 24 feet | 3 7 beads | 4 4 times | 5 20 toy cars | 6 24 coins

Explanations

1 *Multiply Amy's 7 stickers by 6:* $7 \times 6 = 42$ *stickers.*

2 *The python is 6 groups of 4 feet:* $6 \times 4 = 24$ *feet.*

3 *Divide the larger amount by the comparison factor:* $56 \div 8 = 7$ *beads.*

4 *Compare weights by division:* $36 \div 9 = 4$*, so the dog weighs 4 times as much.*

5 *Sister has* $3(5) = 15$ *cars; add Ryan's 5 to get* $15 + 5 = 20$ *cars altogether.*

6 *Four times 8 is 32 total coins; subtract the original 8:* $32 - 8 = 24$ *new coins.*

> Section 1.3

1 26 cookies | 2 16 apples | 3 \$49 | 4 5 vans | 5 12 muffins

6 $s = (4 \times 9) - 10 = 26$

Explanations

Get Online
Find more at
ViewMath.com/TX-Grade4
ViewMath.com

1. Find the total, then subtract what was eaten: $5(6) - 4 = 30 - 4 = 26$ cookies.

2. Each bag has $48 \div 6 = 8$ apples; two bags contain $2(8) = 16$ apples.

3. Total hours are $4 + 3 = 7$; multiply by \$7: $7(7) = \$49$.

4. Since $30 \div 7 = 4$ R2, four vans are not enough; round up to 5 vans.

5. Three sold boxes contain $3(8) = 24$ muffins; $36 - 24 = 12$ remain.

6. Four packs contain $4(9) = 36$ stickers; after giving away 10, $s = 36 - 10 = 26$.

> Section 1.4

1. $1, 2, 3, 6, 9, 18$
2. $(1, 20), (2, 10), (4, 5)$
3. $1, 2, 4, 8, 16, 32$
4. Yes
5. No
6. $1, 2, 4, 7, 14,$ or 28 friends

Explanations

1. List every divisor that makes $18 \div n$ a whole number; this gives $1, 2, 3, 6, 9, 18$.

2. Multiply each listed pair; every product equals 20, so the pairs include all factors.

3. List every divisor that makes $32 \div n$ a whole number; this gives $1, 2, 4, 8, 16, 32$.

4. Check divisibility: $42 \div 7$ is a whole number, so 7 is a factor of 42.

Get Online
Find more at
ViewMath.com/TX-Grade4
ViewMath.com

5 *Check divisibility:* $27 \div 5$ *is not a whole number, so* 5 *is not a factor of* 27.

6 *List every divisor that makes* $28 \div n$ *a whole number, this gives* $1, 2, 4, 7, 14$, *or* 28 *friends.*

Section 1.5

1 $6, 12, 18, 24, 30$ 2 *No* 3 *Prime* 4 *Composite* 5 *False* 6 $11, 13, 17, 19$

Explanations

1 *Multiply* 6 *by* $1, 2, 3, 4, 5$ *to generate the listed multiples:* $6, 12, 18, 24, 30$.

2 *Since* $14 \div 3 = 4$ *remainder* 2, 14 *is not a multiple of* 3.

3 *The only factors of* 17 *are* 1 *and* 17, *so* 17 *is prime.*

4 *Because* 27 *has a factor besides* 1 *and itself, it is composite.*

5 *Prime numbers have exactly two factors;* 1 *has only one factor, so the statement is false.*

6 *Test divisibility by smaller primes; only* $11, 13, 17, 19$ *have exactly two factors in this interval.*

Section 1.6

1 $3, 7, 11, 15, 19, 23$ 2 *Add* 5 3 $4, 2$ 4 *The numbers alternate odd, even* $(1, 4, 7, 10, \ldots)$

5 $28, 35$ 6 $100, 91, 82, 73, 64$

Get Online
Find more at
ViewMath.com/TX-Grade4
ViewMath.com

Explanations

1. *Begin at* 3 *and add* 4 *repeatedly:* $3, 7, 11, 15, 19, 23$.

2. *Compare consecutive terms; each rises by the same amount, so the rule is Add* 5.

3. *Each term is half the preceding term:* $64, 32, 16, 8, 4, 2$.

4. *Adding* 3 *switches parity each time:* $1, 4, 7, 10, \ldots$*, so odd and even terms alternate.*

5. *Continue adding* 7 *after* 21*:* $21 + 7 = 28$ *and* $28 + 7 = 35$.

6. *Begin at* 100 *and subtract* 9 *repeatedly:* $100, 91, 82, 73, 64$.

Chapter 2

Section 2.1

1. Ten-thousands
2. 70,000,000
3. 9,000,000,000
4. Six hundred five thousand, ninety
5. 3,200,000,000
6. 8,000,000,000

Explanations

1. *Read place values from the right; the named digit is in the ten-thousands place.*

2 *Multiply the digit by its place value in the numeral; its value is* 70,000,000.

3 *Multiply the digit by its place value in the numeral; its value is* 9,000,000,000.

4 *Read each three-digit period from left to right: Six hundred five thousand, ninety.*

5 *Place each named period in its correct positions; the standard form is* 3,200,000,000.

6 *Multiply the digit by its place value in the numeral; its value is* 8,000,000,000.

> Section 2.2

1 $6{,}000 + 200 + 30 + 1$

2 $300{,}000 + 40{,}000 + 900 + 10 + 5$

3 $800{,}000 + 2{,}000 + 60$

4 509,320

5 470,008

6 8,000

Explanations

1 *Write each nonzero digit by its place value, giving* $6{,}000 + 200 + 30 + 1$.

2 *Write each nonzero digit by its place value, giving* $300{,}000 + 40{,}000 + 900 + 10 + 5$.

3 *Write each nonzero digit by its place value, giving* $800{,}000 + 2{,}000 + 60$.

4 *Add the expanded place-value parts to form the standard number* 509,320.

5 *Add the expanded place-value parts to form the standard number* 470,008.

6 *The missing addend supplies the thousands place, so it is* $8{,}000$.

> *Section 2.3*

1 $>$	2 $<$	3 $>$	4 $=$	5 $7{,}500 < 75{,}000 < 750{,}000 < 7{,}500{,}000$	6 *California*

Explanations

1 *Compare digits from the greatest place; the first difference shows* $67{,}432 > 67{,}234$.

2 *Compare digits from the greatest place; the first difference shows* $5{,}410{,}000 < 5{,}420{,}000$.

3 *Compare digits from the greatest place; the first difference shows* $3{,}000{,}000{,}000 > 2{,}999{,}999{,}999$.

4 *Compare digits from the greatest place; the first difference shows* $83{,}450{,}000 = 83{,}450{,}000$.

5 *Compare place values from greatest to least; the ascending order is* $7{,}500 < 75{,}000 < 750{,}000 < 7{,}500{,}000$.

6 *Compare the first place where the populations differ; California has the larger value.*

> *Section 2.4*

1 340	2 5,600	3 420	4 9,000	5 100 *times*	6 10

Explanations

Get Online
Find more at
ViewMath.com/TX-Grade4
ViewMath.com

1 *Multiplying by* 10 *shifts every digit one place left:* $34 \times 10 = 340$.

2 *Multiplying by* 10 *shifts every digit one place left:* $560 \times 10 = 5{,}600$.

3 *Dividing by* 10 *shifts every digit one place right:* $4{,}200 \div 10 = 420$.

4 *Dividing by* 10 *shifts every digit one place right:* $90{,}000 \div 10 = 9{,}000$.

5 *The thousands-place* 6 *is worth* $6{,}000$ *and the tens-place* 6 *is worth* 60; $6{,}000 \div 60 = 100$.

6 *Solve* $n \times 80 = 800$ *by dividing:* $n = 800 \div 80 = 10$.

Section 2.5

1 300 | 2 $68{,}000$ | 3 $500{,}000$ | 4 $7{,}000{,}000$ | 5 $6{,}000{,}000{,}000$ | 6 *Nearest million*

Explanations

1 *Check the digit immediately right of the target place and round to* 300.

2 *Check the digit immediately right of the target place and round to* $68{,}000$.

3 *Check the digit immediately right of the target place and round to* $500{,}000$.

4 *Check the digit immediately right of the target place and round to* $7{,}000{,}000$.

5 *Check the digit immediately right of the target place and round to* $6{,}000{,}000{,}000$.

6 *The estimate changes the number to the stated million or thousand benchmark, so it was rounded to the nearest million.*

Chapter 3

Section 3.1

1 885 | 2 887 | 3 5,843 | 4 6,572 | 5 2,733 *cans* | 6 6,240 *books*

Explanations

1 *Align place values and regroup as needed:* $524 + 361 = 885$.

2 *Align place values and regroup as needed:* $739 + 148 = 887$.

3 *Align place values and regroup as needed:* $2{,}456 + 3{,}387 = 5{,}843$.

4 *Align place values and regroup as needed:* $4{,}809 + 1{,}763 = 6{,}572$.

5 *Add the two quantities by place value:* $1,258 + 1,475 = 2{,}733$.

6 *Add the two quantities by place value:* $3,692 + 2,548 = 6{,}240$.

Section 3.2

1 444 | 2 326 | 3 3,438 | 4 4,537 | 5 2,225 *apples* | 6 2,355 *fewer visitors*

Explanations

1. *Align place values and regroup as needed:* $876 - 432 = 444$.

2. *Align place values and regroup as needed:* $615 - 289 = 326$.

3. *Align place values and regroup as needed:* $5{,}274 - 1{,}836 = 3{,}438$.

4. *Align place values and regroup as needed:* $7{,}005 - 2{,}468 = 4{,}537$.

5. *Subtract the amount removed from the starting total:* $3,500 - 1,275 = 2{,}225$.

6. *Subtract the amount removed from the starting total:* $8,042 - 5,687 = 2{,}355$.

Section 3.3

1. 498 2. 1,235 3. 4,952 4. 12,645 5. 2,688 *books* 6. 7,500 *square feet*

Explanations

1. *Use place-value partial products, then add them:* $6 \times 83 = 498$.

2. *Use place-value partial products, then add them:* $5 \times 247 = 1{,}235$.

3. *Use place-value partial products, then add them:* $8 \times 619 = 4{,}952$.

4. *Use place-value partial products, then add them:* $3 \times 4{,}215 = 12{,}645$.

5. *Multiply the number of equal groups by the amount in each:* $7 \times 384 = 2{,}688$.

6. *Multiply the number of equal groups by the amount in each:* $1,250 \times 6 = 7{,}500$.

Section 3.4

1. 294 2. 910 3. 2,544 4. 2,613 5. 432 *pencils* 6. \$900

Explanations

1. *Use place-value partial products, then add them:* $21 \times 14 = 294$.

2. *Use place-value partial products, then add them:* $35 \times 26 = 910$.

3. *Use place-value partial products, then add them:* $48 \times 53 = 2{,}544$.

4. *Use place-value partial products, then add them:* $67 \times 39 = 2{,}613$.

5. *Multiply the number of equal groups by the amount in each:* $24 \times 18 = 432$.

6. *Multiply the number of equal groups by the amount in each:* $12 \times 75 = \$900$.

Section 3.5

1. 24 2. 58 *R*1 3. 81 4. 219 *R*6 5. 41 *stickers per bag,* 4 *left over* 6. 5 *buses*

Explanations

1. *Divide and check with multiplication:* $4 \times 24 = 96$.

2. *Divide and check:* $3(58) + 1 = 175$, *so the quotient is* 58 *R*1.

3. *Divide and check with multiplication:* $8 \times 81 = 648$.

4. *Divide and check:* $7(219) + 6 = 1,539$, *so the quotient is* 219 *R*6.

5. *Divide* 250 *items among groups of* 6*; the quotient and remainder are* 41 *stickers per bag,* 4 *left over.*

6. *Divide students by capacity; because a remainder still needs a seat, round the quotient up to* 5 *buses.*

Section 3.6

1. $500 + 300 = 800$
2. $7{,}000 - 3{,}000 = 4{,}000$
3. $50 \times 60 = 3{,}000$
4. $250 \div 5 = 50$
5. *No.* $200 + 600 = 800$, *so* 1,125 *is not reasonable.*
6. $30 \times 40 = 1{,}200$ *muffins*

Explanations

1. *Round* 493 *to* 500 *and* 318 *to* 300*; then* $500 + 300 = 800$.

2. *Round* 7,215 *to* 7,000 *and* 2,893 *to* 3,000*; the difference is* 4,000.

3. *Round* 47 *to* 50 *and* 63 *to* 60*; then* $50(60) = 3{,}000$.

4. *Use nearby compatible* 250*, which divides evenly:* $250 \div 5 = 50$.

5 Round to benchmark values and compare with the claimed sum: No. $200 + 600 = 800$, so 1,125 is not reasonable.

6 Round 28 to 30 and 36 to 40; about $30(40) = 1{,}200$ muffins.

Chapter 4

Section 4.1

1 2 | 2 9 | 3 3 | 4 $\frac{1}{3}$ | 5 $\frac{4}{6}$ and $\frac{6}{9}$ (or any correct equivalents) | 6 $\frac{6}{8}$ (6 eighths)

Explanations

1 Scale numerator and denominator by the same factor, the missing value is 2.

2 Scale numerator and denominator by the same factor, the missing value is 9.

3 Scale numerator and denominator by the same factor, the missing value is 3.

4 Divide numerator and denominator by their common factor to get $\frac{1}{3}$.

5 Multiply numerator and denominator by the same number, examples are $\frac{4}{6}$ and $\frac{6}{9}$ (or any correct equivalents).

6 Multiply numerator and denominator of $\frac{3}{4}$ by 2: $\frac{3}{4} = \frac{6}{8}$, so Maya needs 6 eighths.

Section 4.2

Get Online
Find more at ViewMath.com/TX-Grade4
ViewMath.com

1 $<$ 2 $>$ 3 $=$ 4 $<$ 5 $\frac{3}{8}, \frac{1}{2}, \frac{3}{4}$ 6 Lily ($\frac{2}{5} = \frac{4}{10} > \frac{3}{10}$)

Explanations

1 *Use common denominators or cross-products to compare $\frac{3}{5}$ and $\frac{4}{5}$; the relation is $<$.*

2 *Use common denominators or cross-products to compare $\frac{2}{3}$ and $\frac{2}{6}$; the relation is $>$.*

3 *Use common denominators or cross-products to compare $\frac{4}{8}$ and $\frac{1}{2}$; the relation is $=$.*

4 *Use common denominators or cross-products to compare $\frac{3}{4}$ and $\frac{5}{6}$; the relation is $<$.*

5 *Rewrite with comparable denominators or decimals; the least-to-greatest order is $\frac{3}{8}, \frac{1}{2}, \frac{3}{4}$.*

6 *Rewrite $\frac{2}{5}$ as $\frac{4}{10}$; because $\frac{4}{10} > \frac{3}{10}$, Lily ate more.*

Section 4.3

1 $\frac{1}{5} + \frac{1}{5} + \frac{1}{5} + \frac{1}{5}$ 2 $\frac{2}{8} + \frac{3}{8}$ and $\frac{1}{8} + \frac{4}{8}$ (or any valid sums) 3 $\frac{5}{6}$ 4 $\frac{3}{4}$

5 False — $\frac{3}{5} + \frac{3}{5} = \frac{6}{5}$, not $\frac{6}{10}$ 6 $\frac{4}{8}$ (or $\frac{1}{2}$)

Explanations

1 *Split the numerator into addends while keeping the denominator unchanged; one valid result is $\frac{1}{5} + \frac{1}{5} + \frac{1}{5} + \frac{1}{5}$.*

2 *Split the numerator into addends while keeping the denominator unchanged; one valid result is $\frac{2}{8} + \frac{3}{8}$ and $\frac{1}{8} + \frac{4}{8}$ (or any valid sums).*

3 *Combine equal-size parts by adding numerators:* $\frac{2}{6} + \frac{3}{6} = \frac{5}{6}$.

4 *Combine equal-size parts by adding numerators:* $\frac{1}{4} + \frac{1}{4} + \frac{1}{4} = \frac{3}{4}$.

5 *Add numerators and keep denominator* 5*:* $\frac{3}{5} + \frac{3}{5} = \frac{6}{5}$, *not* $\frac{6}{10}$, *so the statement is false.*

6 *Subtract equal-denominator fractions:* $\frac{6}{8} - \frac{2}{8} = \frac{4}{8} = \frac{1}{2}$.

Section 4.4

1 $\frac{3}{5}$ | 2 $\frac{8}{10}$ or $\frac{4}{5}$ | 3 $\frac{3}{6}$ or $\frac{1}{2}$ | 4 $\frac{4}{12}$ or $\frac{1}{3}$ | 5 $\frac{6}{8}$ or $\frac{3}{4}$ | 6 $\frac{7}{10}$ mile

Explanations

1 *Keep the common denominator and combine numerators:* $\frac{2}{5} + \frac{1}{5} = \frac{3}{5}$.

2 *Keep the common denominator and combine numerators:* $\frac{3}{10} + \frac{5}{10} = \frac{8}{10}$.

3 *Keep the common denominator and combine numerators:* $\frac{5}{6} - \frac{2}{6} = \frac{3}{6}$.

4 *Keep the common denominator and combine numerators:* $\frac{9}{12} - \frac{5}{12} = \frac{4}{12}$.

5 *Keep the common denominator and combine numerators:* $\frac{2}{8} + \frac{3}{8} + \frac{1}{8} = \frac{6}{8}$.

6 *Add the distances with denominator* 10*:* $\frac{3}{10} + \frac{4}{10} = \frac{7}{10}$ *mile.*

Section 4.5

1 $4\frac{3}{5}$ | 2 $7\frac{2}{8}$ or $7\frac{1}{4}$ | 3 $3\frac{2}{4}$ or $3\frac{1}{2}$ | 4 $2\frac{2}{3}$ | 5 $3\frac{2}{5}$ | 6 $4\frac{2}{4}$ feet or $4\frac{1}{2}$ feet

Explanations

1 *Combine whole and fractional parts, regrouping if needed:* $1\frac{2}{5}+3\frac{1}{5}=4\frac{3}{5}$.

2 *Combine whole and fractional parts, regrouping if needed:* $2\frac{3}{8}+4\frac{7}{8}=7\frac{2}{8}$.

3 *Combine whole and fractional parts, regrouping if needed:* $5\frac{3}{4}-2\frac{1}{4}=3\frac{2}{4}$.

4 *Combine whole and fractional parts, regrouping if needed:* $6\frac{1}{3}-3\frac{2}{3}=2\frac{2}{3}$.

5 *Divide* 17 *by* 5*:* $17=5(3)+2$, *so* $\frac{17}{5}=3\frac{2}{5}$.

6 *Regroup and subtract:* $8\frac{1}{4}-3\frac{3}{4}=7\frac{5}{4}-3\frac{3}{4}=4\frac{2}{4}=4\frac{1}{2}$ *feet.*

Section 4.6

1 $\frac{6}{5}$ or $1\frac{1}{5}$ | 2 $\frac{5}{3}$ or $1\frac{2}{3}$ | 3 $\frac{20}{8}$ or $2\frac{4}{8}$ or $2\frac{1}{2}$ | 4 $\frac{14}{6}$ or $2\frac{2}{6}$ or $2\frac{1}{3}$

5 $\frac{1}{4}+\frac{1}{4}+\frac{1}{4}+\frac{1}{4}+\frac{1}{4}=\frac{5}{4}=1\frac{1}{4}$ | 6 $\frac{12}{3}=4$ feet

Explanations

1 *Multiply the whole number by the numerator and keep the denominator.* $3\times\frac{2}{5}=\frac{6}{5}$.

2 *Multiply the whole number by the numerator and keep the denominator.* $5\times\frac{1}{3}=\frac{5}{3}$.

3. *Multiply the whole number by the numerator and keep the denominator.* $4 \times \frac{5}{8} = \frac{20}{8}$.

4. *Multiply the whole number by the numerator and keep the denominator.* $7 \times \frac{2}{6} = \frac{14}{6}$.

5. *Add five copies of* $\frac{1}{4}$: $\frac{1}{4} + \frac{1}{4} + \frac{1}{4} + \frac{1}{4} + \frac{1}{4} = \frac{5}{4} = 1\frac{1}{4}$.

6. *Multiply distance per minute by* 6: $6(\frac{2}{3}) = \frac{12}{3} = 4$ *feet.*

Section 4.7

1. 20
2. $\frac{58}{100}$
3. $\frac{75}{100}$
4. $\frac{60}{100}$ *(or* $\frac{6}{10}$*)*
5. 70 *pennies*
6. $\frac{65}{100}$ *of a dollar (*$\$0.65$*)*

Explanations

1. *Rename each amount with the target denominator, then combine:* $\frac{2}{10} = \frac{?}{100} = 20$.

2. *Rename each amount with the target denominator, then combine:* $\frac{5}{10} + \frac{8}{100} = \frac{58}{100}$.

3. *Rename each amount with the target denominator, then combine:* $\frac{3}{10} + \frac{45}{100} = \frac{75}{100}$.

4. *Multiply numerator and denominator to rename the fraction; the missing numerator is* $\frac{60}{100}$ *(or* $\frac{6}{10}$*).*

5. *Seven tenths of a dollar is* $\frac{70}{100}$*, which equals* 70 *pennies.*

6. *Rewrite tenths as hundredths or thousandths, then add the numerators to get* $\frac{65}{100}$ *of a dollar (*$\$0.65$*).*

Section 4.8

1. 0.6 2. 0.42 3. 0.07 4. 9 5. $\frac{45}{100}$ 6. $0.80 = \$0.80$

Explanations

1. *Use the denominator for decimal place value:* $\frac{6}{10} = 0.6$.
2. *Use the denominator for decimal place value:* $\frac{42}{100} = 0.42$.
3. *Use the denominator for decimal place value:* $\frac{7}{100} = 0.07$.
4. *Use the denominator for decimal place value:* $0.9 = \frac{?}{10} = 9$.
5. *Use hundredths for cents: the decimal represents* $\frac{45}{100}$ *of one dollar.*
6. *Eighty hundredths is* 0.80*, which is the dollar amount* $\$0.80$.

> Section 4.9

1. $=$ 2. $<$ 3. $>$ 4. $<$ 5. $\$1.05, \$1.35, \$1.50, \$1.55$ 6. *Ava*

Explanations

1. *Append trailing zeros, then compare place values:* $0.6 = 0.60$.
2. *Append trailing zeros, then compare place values:* $0.35 < 0.4$.
3. *Compare hundredths after the equal* 2 *ones:* $\$2.75 > \$2.57$ *because* $7 > 5$ *tenths.*

4 *Nine hundredths is less than ten hundredths:* $\$0.09 < \$0.10$.

5 *Write equal numbers of decimal places, then compare digits; the order is* $\$1.05, \$1.35, \$1.50, \$1.55$.

6 *Compare tenths first:* $\$0.54 > \$0.45$, *so Ava has more money.*

> *Section 4.10*

1	8.87	2	12.24	3	6.33	4	5.65	5	10.28	6	\$5.23

Explanations

1 *Align decimal points and combine by place value:* $3.24 + 5.63 = 8.87$.

2 *Align decimal points and combine by place value:* $7.85 + 4.39 = 12.24$.

3 *Align decimal points and combine by place value:* $9.54 - 3.21 = 6.33$.

4 *Align decimal points and combine by place value:* $10.00 - 4.35 = 5.65$.

5 *Align decimal points and combine by place value:* $6.5 + 3.78 = 10.28$.

6 *Align decimal points and combine the money or measurement amounts to get* $\$5.23$.

Chapter 5

Get Online
Find more at
ViewMath.com/TX-Grade4
ViewMath.com

Section 5.1

1. 700 2. 4,000 3. 9,000 4. 2,000 5. 6,000 6. 1,500 *mL*

Explanations

1. *One m equals* 100 *cm, so* $7 \times 100 = 700$ *centimeters.*

2. *One km equals* $1,000$ *m, so* $4 \times 1,000 = 4,000$ *meters.*

3. *One kg equals* $1,000$ *g, so* $9 \times 1,000 = 9,000$ *grams.*

4. *One L equals* $1,000$ *mL, so* $2 \times 1,000 = 2,000$ *milliliters.*

5. *One m equals* $1,000$ *mm, so* $6 \times 1,000 = 6,000$ *millimeters.*

6. *Multiply* 500 *mL per bottle by* 3: $500(3) = 1{,}500$ *mL.*

Section 5.2

1. 48 2. 6 3. 48 4. 20 5. 14 6. 8 *cups* = 4 *pints*

Explanations

1. *Each ft contains* 12 *in, so* $4 \times 12 = 48$ *inches.*

2. *Each yd contains* 3 *ft, so* $2 \times 3 = 6$ *feet.*

Get Online
Find more at
ViewMath.com/TX-Grade4
ViewMath.com

3. *Each lb contains* 16 *oz, so* $3 \times 16 = 48$ *ounces.*

4. *Each gal contains* 4 *qt, so* $5 \times 4 = 20$ *quarts.*

5. *Each qt contains* 2 *pt, so* $7 \times 2 = 14$ *pints.*

6. *Four batches need* $2(4) = 8$ *cups; since* 2 *cups make a pint,* $8 \div 2 = 4$ *pints.*

> *Section 5.3*

1. 180 *minutes*
2. 45 *minutes*
3. $\$3.75$
4. $\$2.65$
5. $5{:}50$ *PM*
6. 250 *cm* (5 *m* $= 500$ *cm;* $500 - 250 = 250$)

Explanations

1. *Each hour has* 60 *minutes, so* $3 \times 60 = 180$ *minutes.*

2. *From* $9{:}15$ *to* $10{:}00$ *is* 45 *minutes.*

3. *Align the decimal points and add both prices:* $\$1.50 + \$2.25 = \$3.75$.

4. *Subtract the purchase total from the payment:* $\$10.00 - \$7.35 = \$2.65$.

5. *Add* 1 *hour* 20 *minutes to* $4{:}30$ *PM to reach* $5{:}50$ *PM.*

6. *Convert* 5 *m to* 500 *cm, then subtract what Sam used:* $500 - 250 = 250$ *cm.*

Get Online
Find more at
ViewMath.com/TX-Grade4
ViewMath.com

> Section 5.4

1 28 sq in | 2 60 sq cm | 3 81 sq ft | 4 45 sq m | 5 4 cm | 6 5 bags

Explanations

1 Use $A = l \times w$: $7 \times 4 = 28$ square units.

2 Use $A = l \times w$: $12 \times 5 = 60$ square units.

3 Use $A = l \times w$: $9 \times 9 = 81$ square units.

4 Use $A = l \times w$: $15 \times 3 = 45$ square units.

5 Use $A = lw$ and divide by length: $w = 36 \div 9 = 4$ cm.

6 Garden area is $10(6) = 60$ ft^2; divide by 12 ft^2 per bag to get 5 bags.

> Section 5.5

1 26 ft | 2 36 cm | 3 24 m | 4 6 in | 5 False | 6 46 ft

Explanations

1 Use $P = 2l + 2w$: $2(9) + 2(4) = 26$.

2 Use $P = 2l + 2w$: $2(11) + 2(7) = 36$.

3 Use $P = 2l + 2w$: $2(6) + 2(6) = 24$.

4 Half the perimeter is $40 \div 2 = 20$; subtract the known side: $20 - 14 = 6$.

5 A square with side 5 has perimeter $4(5) = 20$ cm, not 25 cm, so the statement is false.

6 Add all four garden sides with $P = 2l + 2w$; the required fencing is 46 ft.

> Section 5.6

1 3 | 2 8 | 3 $\frac{2}{8}$ in | 4 $\frac{4}{8}$ or $\frac{1}{2}$ in | 5 $\frac{4}{8} + \frac{5}{8} = \frac{9}{8} = 1\frac{1}{8}$ in

Explanations

1 Count the plotted marks that satisfy the question; the count is 3.

2 Count the plotted marks that satisfy the question; the count is 8.

3 The tallest stack of marks is above $\frac{2}{8}$ in, so that value is the mode.

4 Subtract the smallest plotted value from the largest; the range is $\frac{4}{8}$ or $\frac{1}{2}$ in.

5 Add the two greatest plotted measurements: $\frac{4}{8} + \frac{5}{8} = \frac{9}{8} = 1\frac{1}{8}$ in.

> Section 5.7

1 360° | 2 90° | 3 4 | 4 True | 5 90° | 6 90°

Get Online
Find more at
ViewMath.com/TX-Grade4
ViewMath.com

Explanations

1. A complete rotation measures 360°, so a full turn is 360°.

2. A right angle is one quarter of 360°: $360 \div 4 = 90^\circ$.

3. Each right angle is 90°; $360 \div 90 = 4$ right angles make a full turn.

4. A straight angle is half of a full turn: $360^\circ \div 2 = 180^\circ$, so the statement is true.

5. Four equal slices divide the 360° center evenly: $360 \div 4 = 90^\circ$ each.

6. One fourth of a full 360° turn is $360 \div 4 = 90^\circ$.

> Section 5.8

1 Acute | 2 Obtuse | 3 False | 4 90° | 5 180° | 6 Jada (150°)

Explanations

1. An acute angle measures more than 0° but less than 90°, so this angle is acute.

2. An obtuse angle is greater than 90° and less than 180°, so this angle is obtuse.

3. Check the stated angle fact against the 0° to 180° protractor scale; it is false.

4. Use the protractor scale or the angle definition to identify the result as 90°.

Get Online
Find more at
ViewMath.com/TX-Grade4
ViewMath.com

5. *Use the protractor scale or the angle definition to identify the result as* $180°$.

6. *Use the protractor scale or the angle definition to identify the result as Jada* ($150°$).

> Section 5.9

1	2	3	4	5	6
$90°$	$70°$	$52°$	$90°$	True	$130°$

Explanations

1. *Use angle addition and solve with* $65 + 25$*; the missing or total angle is* $90°$.

2. *Use angle addition and solve with* $180 - 110$*; the missing or total angle is* $70°$.

3. *Use angle addition and solve with* $90 - 38$*; the missing or total angle is* $52°$.

4. *Use angle addition and solve with* $45 + 45$*; the missing or total angle is* $90°$.

5. *Angles totaling* $90°$ *form a right angle, so the statement is true.*

6. *Angles around the center total* $360°$*;* $360 - 230 = 130°$ *for the smaller piece.*

> Section 5.10

1	2	3	4	5	6
Fixed	Variable	$5.00	$27	$2	True

Explanations

Get Online
Find more at
ViewMath.com/TX-Grade4
ViewMath.com

1 *The piano payment stays $\$40$ every month, so it is a fixed expense.*

2 *The grocery total changes from $\$250$ to $\$310$, so it is a variable expense.*

3 *Cost is $10(0.50) = \$5$ and revenue is $10(1.00) = \$10$; profit is $\$10 - \$5 = \$5$.*

4 *Profit equals earnings minus supplies: $\$42 - \$15 = \$27$.*

5 *Subtract savings and spending from the allowance: $\$25 - \$8 - \$10 - \$5 = \$2$.*

6 *Profit equals revenue minus cost and is positive when revenue exceeds spending, so the statement is true.*

> Section 5.11

1 11 | 2 10 *minutes* | 3 23 *minutes* | 4 24 ($34 - 10 = 24$) | 5 *False* | 6 *True*

Explanations

1 *Count all leaves in the stem-and-leaf plot; the 11 leaves represent 11 people.*

2 *The smallest stem-and-leaf value is $1|0$, which represents 10 minutes.*

3 *The value 23 appears most often in the display, so the mode is 23 minutes.*

4 *Subtract the fastest time from the slowest: $34 - 10 = 24$ minutes.*

5 *Leaves in a stem-and-leaf plot are ordered least to greatest, so the statement is false.*

6 *Each dot records one observation at its value, so the statement is true.*

> Section 5.12

1 240 *minutes*

2 5 *minutes*

3 1 *hour* 30 *minutes*

4 5:55 *PM*

5 4:00 *PM*

6 *False*

Explanations

1 *Use* 60 *minutes per hour and multiply; the converted time is* 240 *minutes.*

2 *Divide by* 60 *seconds per minute; the converted time is* 5 *minutes.*

3 *From* 10:15 *to* 11:15 *is one hour, then* 30 *more minutes gives* 1 *hour* 30 *minutes.*

4 *Add* 1 *hour to reach* 5:30 *PM, then* 25 *minutes to reach* 5:55 *PM.*

5 *Central Time is one hour behind Eastern Time, so* 5:00 *PM becomes* 4:00 *PM.*

6 *Two hours equal* $2(60) = 120$ *minutes, not* 100*, so the statement is false.*

Chapter 6

> Section 6.1

1 1 | 2 2 | 3 False | 4 4 | 5 $\overleftrightarrow{MN}$ | 6 5

Explanations

1 *A ray begins at one endpoint and extends forever in one direction, so it has* 1 *endpoint.*

2 *A segment begins and ends at fixed points, so it has* 2 *endpoints.*

3 *Use the defining arrows and endpoints of the named figure; the correct description is False.*

4 *Use the defining arrows and endpoints of the named figure; the correct description is* 4.

5 *Use the defining arrows and endpoints of the named figure; the correct description is* $\overleftrightarrow{MN}$.

6 *Use the defining arrows and endpoints of the named figure; the correct description is* 5.

Section 6.2

1 Acute | 2 Right | 3 Obtuse | 4 Straight | 5 60° and 30° are acute; 90° is a right angle | 6 True

Explanations

1 *An acute angle measures more than* 0° *but less than* 90°*, so this angle is acute.*

2 *A* 90° *angle is defined as a right angle.*

3 *An obtuse angle is greater than 90° and less than 180°, so this angle is obtuse.*

4 *Use the protractor scale or the angle definition to identify the result as Straight.*

5 *Use the protractor scale or the angle definition to identify the result as 60° and 30° are acute; 90° is a right angle.*

6 *Use the protractor scale or the angle definition to identify the result as True.*

> Section 6.3

1 False

2 True

3 2

4 4

5 Railroad tracks, opposite edges of a book, or similar

6 True

Explanations

1 *Parallel lines remain equidistant and never intersect; applying that definition gives False.*

2 *Perpendicular lines meet at 90°, so they form right angles.*

3 *A square's opposite sides form two parallel pairs: top-bottom and left-right.*

4 *Each adjacent side pair meets at one of a square's four right-angle corners, giving 4 pairs.*

5 *Railroad rails stay the same distance apart and do not meet, so they model parallel lines.*

6 *The top and stem of a capital T meet at a 90° angle, so the statement is true.*

Section 6.4

1. Square
2. Trapezoid
3. True
4. Acute triangle
5. 6
6. True

Explanations

1. Four equal sides and four right angles satisfy the definition of a square.

2. A quadrilateral with exactly one pair of parallel sides is a trapezoid.

3. A square has four right angles, so it meets every requirement for a rectangle; the statement is true.

4. All three angles are below $90°$, so the triangle is acute.

5. The prefix "hex-" means six, so a hexagon has 6 sides.

6. A rectangle has two pairs of parallel opposite sides, so every rectangle is a parallelogram.

Section 6.5

1. 3
2. 2
3. False
4. 5
5. Yes, 1 (vertical)
6. A vertical line down the middle

Explanations

1. Each vertex-to-opposite-midpoint fold matches the halves, giving an equilateral triangle 3 symmetry lines.

2. A nonsquare rectangle matches across its vertical and horizontal midlines, giving 2 symmetry lines.

3 *Every diameter is a symmetry line, so a circle has infinitely many, not exactly 4.*

4 *A regular pentagon has one symmetry line through each of its 5 vertices and the opposite side midpoint.*

5 *A vertical fold through a capital A matches its left and right halves, giving 1 symmetry line.*

6 *A vertical fold through the heart's center makes its left and right halves coincide.*

Great job checking your work!

Keep practicing and you'll be a math star!

Author's Final Note

I hope you enjoyed this book as much as I enjoyed writing it. Whether you are a student working through the material, a parent supporting your child's learning, or a teacher guiding your class, I have tried to make this book as clear and engaging as possible. I hope I have succeeded. If you have any suggestions for improvement, please let me know. I would love to hear from you.

The accuracy of calculations is very important to me. We have done our best, but I also expect that I have made some minor errors. Constant improvement is the name of the game. If you find any errors, please let me know. I will fix them in the next edition.

For students: *Your learning journey does not end here. I have written a series of books to help you learn math. Make sure you browse through them. I especially recommend workbooks and practice tests to help you prepare for your exams.*

For parents: *Thank you for investing in your child's education. I encourage you to explore the companion resources available online to help support your child outside the classroom.*

For teachers: *Thank you for the invaluable work you do every day. I hope this book serves as a useful resource in your classroom. Feel free to reach out if you have suggestions or would like to discuss how best to use this book with your students.*

I also enjoy reading your reviews. If you have a moment, please leave a review on where you found this book. It will help others find this book. If you have any questions or comments, please feel free to contact me at dr.nazari@ViewMath.com.

And one last thing: Remember to use online resources for additional help. I recommend using the resources on `https://ViewMath.com` *You can find video lessons, practice problems, and more. You can also use the online companion for this book to track your progress and access additional resources.*

Wishing all students the best in their studies, parents every success in supporting their children, and teachers continued inspiration in their classrooms!

Dr. A. Nazari

www.ingramcontent.com/pod-product-compliance
Lightning Source LLC
LaVergne TN
LVHW082336200926
848481LV00012B/3358